# From Hiroshima to the Iceman
## The Development and Applications of
## Accelerator Mass Spectrometry

## Also by Harry E Gove

### Relic, Icon or Hoax?
### Carbon Dating the Turin Shroud

Published in 1996, *Relic, Icon or Hoax?* tells the story of how the age of the Turin Shroud came to be determined. This is an eyewitness account of the issues that arose before, during and after the carbon dating took place, and details the strange inter-relationships between religion and science that characterized the whole endeavour.

A very personal account from someone who was involved in the process from start to finish.

### Available from Institute of Physics Publishing

# From Hiroshima to the Iceman

## The Development and Applications of Accelerator Mass Spectrometry

**Harry E Gove**

Professor Emeritus of Physics
University of Rochester

Institute of Physics Publishing
Bristol and Philadelphia

*British Library Cataloguing-in-Publication Data*
A catalogue record for this book is available from the British Library.

ISBN  0 7503 0557 6 (hbk)
ISBN  0 7503 0558 4 (pbk)

*Library of Congress Cataloging-in-Publication Data are available*

Published by Institute of Physics Publishing, wholly owned by The Institute of Physics, London

Institute of Physics Publishing, Dirac House, Temple Back, Bristol BS1 6BE, UK
US Office: Institute of Physics Publishing, Suite 1035, The Public Ledger Building, 150 South Independence Mall West, Philadelphia, PA 19106, USA

Typeset in TeX using the IOP Bookmaker Macros
Printed in Great Britain by J W Arrowsmith Ltd, Bristol

To my daughters Pauline and Diana

# Contents

# Preface

Accelerator mass spectrometry (AMS) is the most recent of the many applied and/or non-nuclear spin-offs originating from basic research in nuclear science. The goal of such basic research is, primarily, to understand the properties of the atomic nucleus and the nature of the interactions between nuclei. In the process of achieving that understanding the research has led to many applications, most of them of great benefit to mankind. In addition and prior to AMS these include nuclear power, the production and use of radioisotopes for medicine including their use in imaging human organs, the use of radioisotope tracers in industry, the use of accelerators for cancer therapy and for studies of condensed matter especially of semiconductors, and a myriad of other applications. A more controversial spin-off, of course, is the development of nuclear weapons. Nuclear science has also impacted many other fields of science including particle physics, astrophysics, archaeology, geology and several others.

This book gives an account of the historical development of AMS, the development of tandem electrostatic accelerators, designed for use in basic research in nuclear science that, in addition, are almost exclusively used in the practice of AMS, the instrumentation of these accelerators for AMS, and some of its applications. It is intended to provide potential users of AMS who do not have a background in nuclear physics, as well as people who have a general interest in science, with an understanding of the technique, its power and its limitations.

The applications of AMS increase almost daily. That is, to those who have been involved in its development, the most astonishing hallmark of the technique. As soon as it is demonstrated that a particular isotope can be detected with ultrasensitivity by AMS reasons for doing so follow rapidly. Perhaps the exemplar of this is chlorine-36. This isotope of chlorine is radioactive and has a half-life (the time it takes for half the atoms in a sample of chlorine-36 to decay) of 301 000 years. The original incentive for detecting it at high sensitivity in samples of milligram size was the realization (as described in this book) that it could be used to measure the age of ground water on a million year time scale. This information is crucial for the determination of the suitability of potential sites for the storage of nuclear waste. Shortly after it was shown that chlorine-36 could be detected by AMS at an appropriate sensitivity and the information was widely disseminated other applications were soon suggested and pursued. These included the measurement of the leakage of nuclear waste into the environment from governmental laboratories dedicated to the development and manufacture of nuclear weapons and nuclear reactors, the determination of the numbers of neutrons released in the explosion of nuclear bombs over Hiroshima and Nagasaki, the exposure ages of rocks and hence, for example, the ages of craters created by major meteorite impacts, the terrestrial ages of meteorites incident on Antarctica and the rate of flow of ground water and more. All these are discussed in the book.

For most people, however, the most fascinating aspect of AMS is its ability to determine the ages of organic material, animal and vegetable, in samples weighing a few milligrams or less by the detection of carbon-14, a radioactive isotope of carbon with a half-life of 5730 years. Examples are given in the book of how carbon dating by AMS is providing answers to where, when and by whom the Americas were initially peopled, the origins of agriculture by America's native people, when the Vikings came to North America, the ages of the Dead Sea scrolls, the age of an elephant bird egg found in Australia and when the famous 'Iceman', discovered in 1991, was frozen in an icy grave in the Italian alps.

The future of AMS is secure. The answers it will provide to fascinating riddles in the years ahead will continue to amaze and amuse mankind.

**Harry E Gove**
November 1998

# Acknowledgments

It is a pleasure to acknowledge the contributions of the various laboratories that pioneered the development of accelerator mass spectrometry. These include the *General Ionex/Toronto/Rochester* group led by K H Purser, A E Litherland and the author; other members of this group who made major contributions include C L Bennett, R P Beukens, M R Clover, K H Chang, D Elmore, R Ferraro, B R Fulton, M P Gorton, L R Kilius, H W Lee, R B Liebert, J R Marsden, H Naylor, K Nishiizumi, J C Rucklidge and W E Sondheim. Other laboratories include *Lawrence Berkeley National Laboratory* with R A Muller, E J Stephenson and T S Mast, the *Simon Fraser–McMaster Universities* with D E Nelson, R G Kortelling, D G Burke, J W Mckay, W R Stott, J Southon and J S Vogel, the *Rene Bernas Laboratory* with G M Raisbeck, F Yiou, M Fruneau, M Lieuvin, J C Ravel and J M Loiseaux, the *US Geological Survey* with M Rubin, the *University of California, San Diego* with J R Arnold, K Nishiizumi and R C Finkel, the *Chalk River Nuclear Laboratories of Atomic Energy of Canada Ltd* with H R Andrews, G C Ball, R M Brown, N Burn, W G Davies, Y Imahori and J C D Milton, the *Nuclear Physics Laboratory and the Research Laboratory for Archaeology and the History of Art, Oxford University* with P J S B Barrat, G Doucas, E F Garman, H R McK Hyder, S H Chew, T J L Greenway, K W Allen, D Sinclair, E T Hall, R E M Hedges and N R White, the *University of Pennsylvania* with R Middleton, J Klein and W E Stephens, the *ETH Zurich and Bern University* with M Suter, R Balzer, G Bonani, Ch Stoller, W Woelfli, J Beer, H Oeschger and B Stauffer, the *University of Washington* with G W Farwell, F H Schmidt, M Stuiver and P M Grootes, the *Hebrew University and the Weizmann Institute* with M Paul, R Kaim, A Breskin, R Chechik, J Gerber, M B Golberg, N Zwang and A Kaufman, *Argonne National Laboratory* with J P Schiffer, H Ernst,

W Henning, W Kutschera, B Myslek-Laurikainen, R C Pardo, R K Smither and J L Yntema, the *Technical University, Munich* with P W Kubik, G Korschinek, E Nolte, M S Pravikoff and H Morinaga, the *University of Arizona* with D J Donahue, P E Damon, A Long, A T J Jull, T H Zabel, S N Davis and H Bentley, and *Nagoya University* with M Furukawa, N Nakai and E Nakano.

There are a number of people whose roles I would like to warmly acknowledge as collaborators, advisors or teachers, often all three. These include Tom Beasley, *Environmental Measurements Laboratory (DOE), New York*, Tom Cahill, *Crocker Cyclotron Laboratory, University of California at Davis*, Doug Donahue, *NSF-Arizona AMS Laboratory, University of Arizona*, David Elmore, *PRIME Lab AMS Facility, Purdue University*, Leoncio Garza-Valdes, *San Antonio, Texas*, Vance Haynes, *Departments of Anthropology and Geosciences, University of Arizona*, Tim Jull, *NSF-Arizona AMS Laboratory, University of Arizona*, Ted Litherland, *IsoTrace Laboratory, University of Toronto*, Steve Mattingly, *University of Texas Health Science Center at San Antonio*, Ken Purser, *Southern Cross Corporation, Danvers, Massachusetts*, Meyer Rubin, *US Geological Survey, Reston, Virginia*, Konrad Spindler, *Forschungsinstitut für Alpine Vorzeit, University of Innsbruck, Austria*, Tore Straume, *Lawrence Livermore National Laboratory, University of California*, Erv Taylor, *Radiocarbon Laboratory, Department of Anthropology and Institute of Geophysics and Planetary Physics, University of California at Riverside* and John Vogel, *Center for AMS, Lawrence Livermore National Laboratory, University of California*.

I would also like to thank Ray Teng and David Munson of the Nuclear Structure Research Laboratory and the Department of Physics and Astronomy, University of Rochester for their valuable assistance.

Harry E Gove
November 1998

# Brief Biographical Sketch of the Author

*Harry E Gove*

Harry Gove was born in Niagara Falls, Ontario, Canada. He received his BSc degree in Engineering Physics at Queen's University in Kingston, Ontario, Canada in 1944 and his PhD degree from the Massachusetts Institute of Technology in Nuclear Physics in 1950. He served as Branch Head of Nuclear Physics at Atomic Energy of Canada Ltd, Chalk River, from 1956 to 1963 and in 1963 he was appointed as a Professor of Physics at the University of Rochester. He became a US citizen in 1969. He directed the Nuclear Structure Research Laboratory at the University of Rochester from 1963 to 1988 and was chair of the Department of Physics and Astronomy from 1977 to 1980. He became Professor Emeritus of Physics at the University of Rochester in 1992 and Adjunct Professor of Physics at the University of Toronto in 1997. He was on leave at the Niels Bohr Institute in Copenhagen, Denmark, in 1961–62, at the Centre for Nuclear Research in Strasbourg, France, in 1971–72 and at Worcester College, Oxford, England, in 1983–84 where he was the R T French Visiting Professor. He has served on numerous visiting and advisory committees in the USA and Canada. He was a member of the Board of Trustees of Associated Universities, Inc. from 1978 to 1983. He was the Nuclear Physics Division Associate Editor of *Physical Review Letters* from 1975 to 1979 and Associate Editor of *Annual Reviews of Nuclear and Particle Physics* from 1978 to 1994. In 1980 he was the recipient of the JARI (*Journal of Applied Radiation and Instrumentation*) Award from Pergamon Press, along with A E Litherland and K H Purser, for outstanding contributions to the development of accelerator-based dating techniques. He served from Sublieutenant to Lieutenant in the Royal Canadian Navy from 1944 to 1945. He is the author of one book, *Relic, Icon or Hoax? Carbon Dating the Turin Shroud* and the author or co-author of some 235 papers on experimental nuclear physics and accelerator mass spectrometry in various scientific journals.

# Chapter 1

# Introduction

This book is a semiautobiographical account of the historical development of accelerator mass spectrometry (AMS), the instrumentation involved and some of its more interesting, if not to say recondite, applications. AMS has been used to gain information on when man first arrived in the Americas and to date the discovery of America by the first Europeans, to measure the rate of flow and the age of underground water, to establish the exposure ages of rocks on the rims of meteor craters, and thus reveal the time of the meteorite impact, to measure the blast intensity of neutrons from the Hiroshima and Nagasaki atom bombs, to gain insights on the earth's environmental history, to determine the age of the Turin Shroud [1] and much more. But what is of paramount significance is that it uses samples that are at least a thousand times smaller than those required for any other method. Usually there is no other method.

As will be discussed later, AMS is a technique for the ultrasensitive detection of long-lived radioactive nuclear isotopes of stable atoms. What is a nuclear isotope? The nucleus of any atomic element is made up of positively charged particles called protons and neutral particles called neutrons (particles of almost equal mass to the proton but with no electric charge) packed closely together and bound by the nuclear force. The number of protons in the nucleus determines the number of negatively charged electrons that can be bound in orbits around the nucleus. It is the number of electrons that determine the chemical properties of the atom. But the nucleus of an atom such as carbon with six protons can have, for example, six, seven or eight neutrons. These atoms have atomic masses of 12, 13 and 14 and they are called the

1

isotopes of carbon. The first two are stable but the third, carbon-14 or radiocarbon, is radioactive.

Without doubt the most interesting applications of AMS, at least to the general public, are in the area of radiocarbon dating, and that involves carbon-14. People have an intense interest in the dates of important past events and the ages of important artefacts. Is this small piece of ancient wood really from the true cross? Are the wooden beams found on Mount Ararat in Turkey really from Noah's ark? Could the Turin Shroud really be Christ's burial cloth? Was Columbus the first European to discover America? Measuring the carbon-14 in a sample or radiocarbon dating can provide an answer to all these questions and, in fact, already has.

Often, however, available sample sizes are too small or not enough of an artefact can be sacrificed to get a date by the radiocarbon dating method invented by Willard Libby in the late 1940s [2]. That method, called decay counting, uses grams of material. For example Libby would have needed a piece of cloth from the Turin Shroud the size of a man's handkerchief to carbon date it. No one was going to allow that much of the shroud to be consumed by fire in the process of converting it to the carbon Libby needed to insert into his counters. Or, as another example, there are violins on the market that have internal yellowing labels certifying them to be from the hand of the master violin craftsman Antonio Stradivari of Cremona, Italy, who died in 1737. Dating a Stradivarius violin by the Libby method might be possible, but so much of the instrument would be consumed that it could no longer be played. AMS requires milligrams of material—the Turin Shroud could be dated using a square centimetre of cloth and a violin from a few slivers of wood removed from a non-critical spot. In short the application of AMS to radiocarbon dating means that virtually anything, no matter how small or valuable, can be dated. As with the Libby method, of course, the object must be of organic origin—a product of trees, plants, animals, etc.

Another increasingly important application of AMS lies in the field of biomedical research and, in the foreseeable future, as a diagnostic tool in medicine. Radiocarbon is the most commonly used tracer in biomedical research. Complex hydrocarbons can be labelled at an appropriate site by replacing the stable carbon atom with carbon-14. The way in which a particular hydrocarbon is metabolized in humans and animals can be deduced by measuring the carbon-14 in blood, urine and tissue and even bone marrow samples, as a function of time. AMS permits the radioactive dose of whatever hydrocarbon is to be studied to be reduced to amounts permitted for use in human subjects, and it also reduces the amount of whatever samples are removed for analysis. The field of AMS biomedicine is still in its formative stages but it may well constitute the medical revolution of the future.

Radiocarbon dating, even by the AMS method, is limited at present to objects of organic origin that died 65 000 years ago or less. This limit may, one day, be extended back to 100 000 years but not much further. The limit is, essentially, set by the half-life (the time it takes for half the atoms to decay) of carbon-14, and that is 5730 years. For carbon-14, 60 000 years is ten half-lives or a decrease in the carbon-14 content of a factor of a thousand. To date further back one needs a radioactive element with a much longer half-life. Like radiocarbon, that element must also be produced by cosmic rays, as will be explained later. The next most important cosmogenic radioisotope that fits the bill is chlorine-36. Its half-life is 301 000 years.

AMS can measure chlorine-36 almost as easily as carbon-14 and in equally small quantities. There are, however, not as many dating possibilities for chlorine-36 as for carbon-14. For example, generally it cannot be used to date anything that was once living and growing. It can, however, be used to date the exposure ages of rocks on the earth's surface, the terrestrial and extraterrestrial ages of meteorites and the age of ground water.

Although the latter may seem arcane, it is of great importance. For example, consider the storage of nuclear waste—a problem currently vexing mankind and one that will become increasingly taxing if we continue to resort to nuclear reactors as sources of electrical energy. Although many people oppose the idea of nuclear power it is probably an option we should continue to keep open. At least reactors do not emit carbon dioxide, an atmospheric pollutant that contributes to global warming. The most obvious place to store used fuel elements from nuclear reactors (usually called, incorrectly, atomic reactors) is deep underground in mines or some other place not directly penetrated by ground water. In the case of mines there is bound to be some ground water but if one could show it was hundreds of thousands of years old it would mean it was not mixing with surface water on a short time scale. Any nuclear waste leached out by such deep mine ground water would not return to the biosphere before its radioactivity was reduced to tolerable levels. That, at least, is the hope. AMS measurement of the chlorine-36 content of the water in the mine can be used to reveal its age.

Another potential storage site for used nuclear fuel is in volcanic tuff well above the water table such as at Yucca Mountain in Nevada. At that site the ground water level lies almost 600 m below the surface and the general geographical area around Yucca is quite arid. Caverns carved out of the mountain should be good, dry storage sites. As we shall see, measurements of chlorine-36 can check this assumption.

One can travel back even further in time by AMS measurements of a third cosmogenic radionuclide, iodine-129. This element has a half-life of almost 16 million years. It can be found in ocean sediments and oil deposits. It is also produced in nuclear reactor fuel elements. It is one of the more abundant products of nuclear fission and can be released into the environment as a result of a reactor accident. For example, when the terrible reactor mishap occurred at Chernobyl in 1986 (see Chapter 9) substantial amounts of iodine-129 as well as the much more lethal iodine-131 and other deleterious fission products were distributed over vast areas surrounding the site and throughout the world, with tragic consequences. At two AMS and several other laboratories the iodine-129 content of soil samples collected at Chernobyl has been measured. It provides information on how much of the short-lived cancer-producing iodine-131 was delivered to the inhabitants of the surrounding region resulting in tens of thousands of deaths, particularly to children and pregnant women. The radioactive contamination produced by that reactor explosion continues to be distressingly high and will remain so for many, many years.

The first application of AMS to natural materials occurred in the spring of 1977 in the Nuclear Structure Research Laboratory (NSRL) at the University of Rochester in upstate New York. That laboratory, funded by the National Science Foundation, commenced operations in 1966. Until 1988 the author was director of the NSRL. The laboratory housed a tandem Van de Graaff accelerator designed for research in nuclear physics and, until its decommissioning in 1995, it was one of the top such university laboratories in the USA. Until 1977 it had made only a few modest forays into fields other than nuclear physics. Since then, until the accelerator was shut down, a fifth of its running time had been devoted to the field of AMS.

Following the description of the historical development of accelerator mass spectrometry in Chapters 2 and 3, the development of both single-ended and tandem electrostatic accelerators is described in Chapter 4. A description of how such tandem accelerators are applied to AMS and the instrumentation required follows in Chapter 5. The latter is not a highly technical description of the subject. A much more technical account has been written by Tuniz *et al* [3]. The present book is designed to introduce the technique of AMS to potential users and to the general public.

Chapters 6 to 11 are devoted to some applications of AMS. The applications included by no means constitute a complete set of AMS research areas nor a complete account of a particular application. They are, however, ones of wide appeal. Chapter 12 presents some thoughts on the future of AMS.

# Chapter 2

---

# Historical Development of Accelerator Mass Spectrometry—1977

My involvement with AMS began during the annual Washington meeting of the American Physical Society in April 1977. A nuclear physics colleague of mine dating back to my Chalk River days, A E Litherland, was meeting with K H Purser, president of General Ionex Corporation, in his hotel suite. Litherland and I collaborated in many pioneering experiments at Chalk River, when I was head of the Nuclear Physics Branch there, using the Chalk River tandem. He left Chalk River in 1966 to become a professor of physics at the University of Toronto. Purser is also a nuclear physicist. Before founding the General Ionex Corporation, he was a senior physicist in my laboratory, the Nuclear Structure Research Laboratory, at the University of Rochester, where he was responsible for the installation of the tandem accelerator. He used it for several years in his nuclear physics research programme. He left the university in 1973 to found a company that manufactures tandem components and small tandems for various industrial purposes. Purser and Litherland had been discussing the possibility of measuring carbon-14 with a tandem accelerator when I joined them. I learned later that it was an idea they had independently been thinking about for some time. As mentioned in Chapter 1, carbon-14 is the key to carbon dating.

Carbon dating was invented by Willard Libby in the 1940s [1] and earned him the Nobel Prize for chemistry in 1960. Radioactive carbon-14 is produced in the atmosphere by cosmic rays and, along with the stable isotopes of carbon, it combines with oxygen to form carbon dioxide gas. All living organisms ingest carbon dioxide. Carbon dating involves measuring the ratio of cosmogenic radioactive carbon-14 to the stable carbon isotopes in organic material. As long as plants or animals are alive this ratio of their carbon component is about one part in a trillion and represents the equilibrium between the production of carbon-14 in the atmosphere and its decay. At the time of death of the organic matter the ratio begins to decrease. If death occurred 5730 years ago (the half-life of carbon-14) the ratio is only half a part in a trillion. A measurement of the ratio determines the time of death with considerable precision.

For many years it had been the dream of carbon daters to employ a mass spectrometer that would detect radioactive carbon-14 directly, at the exceedingly low concentration at which it occurs in living organisms, rather than waiting for it to decay as Libby had done. Such a method would reduce the amount of material required to make a dating by several orders of magnitude over the decay counting technique. For example, as early as 1969 Oeschger *et al* recognized this and suggested using mass spectrometric methods for radiocarbon dating [2].

However, all attempts had failed because to a high accuracy carbon-14 weighs the same as the abundant and ubiquitous stable element nitrogen-14. Nuclear physicists familiar with tandem accelerators knew that negative nitrogen ions could not be accelerated through a tandem Van de Graaff. It was necessary to tune the ion source to a negative molecular ion of nitrogen. It was, therefore, assumed that negative nitrogen ions were unstable. If they were the tandem's use of negative ions would instantly solve that most serious interference problem. Traditional mass spectrometers start with positive ions—neutral atoms with one electron removed—and it is as easy to rip one electron from neutral nitrogen as it is from carbon.

Then why not use negative ions in conventional mass spectrometers? The answer is that there is another serious source of mass 14 interference, namely molecular hydrides of stable carbon atoms, and these do form stable negative ions. Such molecules are destroyed if enough electrons are removed in the terminal stripping process.

After I joined the discussions Purser mentioned that he had a patent pending (it had been filed on 1 March 1976) on a tandem accelerator system for the detection of ozone-eating chlorofluorocarbons in the atmosphere that covered the

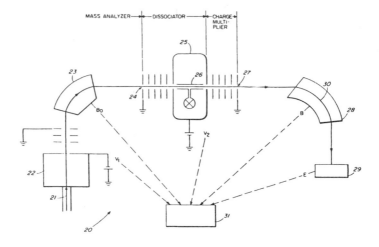

*A diagram of the apparatus proposed by K H Purser in his patent for an ultrasensitive spectrometer for making mass and elemental analyses. Courtesy of K H Purser, Southern Cross Corp., Peabody, MA.*

negative ion source and, most importantly, the principle that molecules accelerated to high velocities would be destroyed in the terminal stripping process [3]. The patent did not, however, specifically address the question of detecting long-lived cosmogenic radioisotopes such as carbon-14.

The questions were just how unstable was the negative nitrogen ion and how many electrons needed to be removed from a carbon hydride molecule to cause it to blow apart? The best way to answer such questions was to make the appropriate measurements on a tandem Van de Graaff accelerator. Litherland and Purser asked whether the Rochester tandem accelerator might be available for this purpose and would I be willing to collaborate. Although this would represent a considerable departure from the basic research in nuclear physics to which my career had been mainly devoted up to then I readily agreed.

I mentioned to Purser and Litherland that at least one other person at Rochester should be involved in the experiments—a graduate student doing his thesis work in the tandem laboratory at Rochester, Charles Bennett. Some time in January Bennett had come into my office at the laboratory to tell me about a violin he had recently purchased for 80 dollars. When he got it home he peered through the sound holes in the body of the instrument and was startled to see a fading, yellow label claiming Stradivari as the maker! He realized that he had either made the most remarkable purchase in the history of violins or that the label was designed to catch the eye. It got him to thinking how he could establish the instrument's age since, if it were a Stradivarius, it would now be almost two and a half centuries old. Violin bodies are, of course, made of wood and as a result contain a reduced carbon-14 to stable carbon ratio appropriate to the date the tree from which the wood came was felled. Carbon dating appeared to be the way to tell whether Bennett's violin was contemporary with Stradivari or not.

In reading up on the subject of carbon dating as practised at that time he quickly realized that, although he could date the violin, he would never be able to play it again. The large amount of wood that would be needed to establish its age by the decay counting method would seriously damage the instrument. We discussed whether it might be possible to take a very small sample of wood, so small as to not affect the instrument's properties, insert it in our tandem's ion source and tune the accelerator system for mass 14. This might directly measure the carbon-14 content of the wood. I considered it an interesting idea but gave it no more thought until my meeting with Purser and Litherland in April.

What Bennett and I did not realize, at the time, was that carbon dating would give an ambiguous answer on organic material that died during the time that Stradivari was building violins in the early eighteenth century. This was the time of the so-called Maunder minimum, a period of 30 years or so when the sun was very quiet. There were no sunspots. During times of a quiet sun the sun's magnetic field is greatly reduced and thus provides less of a shield against the electrically charged cosmic radiation impinging on the earth's atmosphere. The increased intensity of cosmic radiation produces a concomitant increase in carbon-14 production. Trees growing during this time increased their uptake of carbon-14 such that their ratio of carbon-14 to stable carbon is the same as those that grew in the early 1950s.

After the American Physical Society meeting, an article by Richard A Muller of the Lawrence Berkeley Laboratory, California, appeared in the 29 April 1977 issue of *Science* [4]. In this important paper Muller suggested the possibility of using the Berkeley 88 inch positive ion cyclotron to directly detect carbon 14, beryllium-10 and tritium. The carbon-14 would be measured in organic material for the purpose of carbon dating and Muller pointed out that the sample size needed would be substantially less than that required by the Libby decay counting method.

*A photograph of the people involved at the University of Rochester's Nuclear Structure Research Laboratory in the first measurements of carbon-14 in natural samples by AMS in May 1977. They are, from left to right: C L Bennett, A E Litherland, W E Sondheim, H E Gove, R P Clover (below), K H Purser and R P Beukens. The photo was taken by an eighth member of the team, R B Liebert.*

Muller recognized that the main problem would be to separate positive ions of nitrogen-14 from those of carbon-14. He concluded that the easiest and best way of doing so was to exploit the fact that, for the same energy, carbon-14 had a longer path length in an absorber than nitrogen-14. The reason for this is that the rate of energy loss of a charged particle in an absorber is proportional to the square of its charge: carbon has a charge of $6e$ and nitrogen $7e$ ($e$ is the charge on an electron). He felt it would be possible to design an absorber to just stop nitrogen but to transmit carbon-14. Although he had not yet tried to measure carbon-14 in natural samples it was clear that he would soon make the attempt. Muller clearly did not get the idea from us but arrived at it independently. It is rare but not unusual in science, when the time is ripe, for more than one person or group to come up with the same idea independently and virtually simultaneously and that was what happened in this case. In his paper Muller noted that the first use of a cyclotron as a sensitive mass spectrometer was by Alvarez and Cornog in 1939 [5] when they used a 60 inch cyclotron to detect the rare isotope of helium, helium-3, cleanly separating it from tritium. There was no follow-up on the technique, however, until the above paper by Muller and just prior to that an unsuccessful search for new isotopes of hydrogen by Muller *et al*, again using the Berkeley cyclotron, this time the 88 inch [6].

*Apparatus employed for radiofrequency cracking of carbon onto the surface of aluminium cones that are later used in the sputter ion source. H F Gove (left) and R P Beukens (right) are operating the equipment.*

It was some months later that someone pointed out to us that there had been an abstract in the *Bulletin of the American Physical Society* covering the April 1977 meeting of the society authored by Stephenson *et al* [7]. It described some measurements Muller and his colleagues had made on carbon-14 using the Berkeley cyclotron employing samples that contained about a thousand times more carbon-14 than would be found in present-day organic carbon. We had completely overlooked this abstract In that same *Bulletin of the American Physical Society* there was an abstract by Schwarzschild *et al* describing a search they had made for superheavy elements using a tandem accelerator [8].     •

The first AMS experiments on the Rochester tandem took place from 14–20 May 1977. The main aims of the experiments were to determine whether the negative ion of nitrogen was sufficiently stable to reach the tandem's high-voltage terminal (it was not) and how many electrons had to be removed from neutral molecular hydrides of carbon to cause them to completely dissociate (three or more). It was immediately clear that the quest to directly detect carbon-14 in natural organic matter at Rochester would be successful.

We purchased a bag of hardwood barbecue charcoal as an example of present-day carbon since it came from trees that had been felled fairly recently. A milligram or so was inserted in the

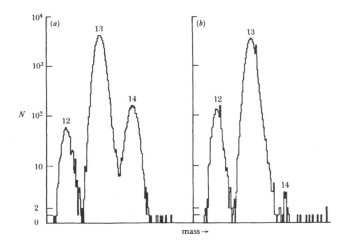

*The first results on the detection of carbon-14 from natural samples by AMS [9, 13]. The data from modern hardwood charcoal is shown in (a) and that from petroleum-derived graphite in (b).*

tandem ion source along with a second sample of graphite as a measure of the background. Graphite is a form of carbon which normally derives from oil and oil deposits are millions of years old. Any carbon-14 with its half-life of 5730 years in graphite derived from oil would have decayed to nothing long ago. On 18 May 1977 the two samples were run sequentially. The charcoal sample gave over a thousand times more carbon-14 counts than did the graphite. We were jubilant. Now the amount of material needed for carbon dating was so small (at least a thousand times smaller than had been required up until then) that it would be possible to date even the most precious artefact. It was one of those instantly recognizable triumphs that occur all too infrequently in science.

Once we had obtained this important result we considered how best to publish it quickly. By tradition first publication in the news media is unacceptable. On the other hand most scientific journals take so long to publish papers that priority, which is important in science, often gets lost. However, a conference on accelerators was to be held in Strasbourg in a few days. Purser had already been invited to give a talk on unrelated apparatus his company was building for a laboratory in Japan. It seemed remarkably timely for him to substitute a paper co-authored by all of us in the group on the AMS work we had just completed. The substitution was made and he delivered the paper at that meeting [9] and it was

*Assembly of the sputter ion source prior to mounting on the Rochester MP tandem. K H Purser (left) and W E Sondheim (right) are inspecting the source.*

to an appropriate scientific audience—a prerequisite to any press release.

However, because the proceedings of a conference like the one in Strasbourg reach a very specialized scientific community, we discussed which journal covering a much broader scientific spectrum would be the most appropriate for us to submit a more carefully crafted paper to. The journals *Science* in the USA and *Nature* in Britain are both prestigious and both cater to a readership covering a wide range of science. We settled on *Science* and began work on the paper.

Our second run took place during early June 1977. We repeated our previous measurements with even better results. At this stage we decided to approach the University of Rochester's Public Relations Office with the idea of issuing a press release. It was sent to the usual news agencies on 8 June 1977. It stressed the fact that the new technique, reported at a scientific meeting in Strasbourg, represented an important breakthrough not only by reducing the sample size by several orders of magnitude but also in allowing much more ancient artefacts to be dated. It turned out that the former was the more important feature of the method. The release was picked up by a number of news organizations including the *New York Times* and an article written by one of their senior science reporters, Boyce Rensberger, was published in the 9 June 1977 edition.

# The New York Times

NEW YORK, THURSDAY, JUNE 9, 1977

## A New Method of Carbon-14 Dating Expected to Double Science's Range

### By BOYCE RENSBERGER

Atomic scientists have devised a new method of carrying out carbon-14 dating of archeological and paleontological specimens that promises to more than double the time span from which ancient organic objects can be dated.

The new method is also said to be much more accurate and to work on samples so tiny that they could not previously be dated because the conventional carbon-14 method would destroy them.

Although the new method has not yet been applied to specimens of scientific significance, the validity of its underlying theory has been confirmed experimentally.

Specialists in studies of the past are typically extremely cautious in accepting new dating methods, but if this new technique wins favor among them, it may lead to a broad wave of new dates on ancient objects. Such a development would certainly add to, and might possibly transform, the understanding of the last 100,000 years on earth.

Conventional carbon-14, or radiocarbon, dating which is widely accepted as reliable, cannot give readings on objects that are more than about 40,000 years old. The new method is said to be useful on objects up to and possibly beyond 100,000 years of age.

The conventional method requires between 10 and 100 grams, or up to a quarter of a pound, of the object to be destroyed in the dating process. This is more than many samples weigh. The new method destroys only 10 to 20 milligrams of a sample, or about seven one-thousandths of an ounce. Such an amount can easily be removed from many samples without destroying their usefulness for other purposes.

#### Mass Spectroscopy Used

The new method which uses techniques of mass spectroscopy, was developed by a team of researchers at the University of Rochester's Nuclear Structure Research Laboratory, which is supported by the National Science Foundation, in collaboration with the General Ionex Corporation of Boston and scientists at the University of Toronto.

"In the past, scientists have determined the age of objects by measuring their carbon-14 radioactivity," said Harry Gove, director of the Rochester laboratory, and a co-developer of the new method along with Dr. Purser and A. E. Litherland of Toronto. "This is like waiting for a clock to tick in order to determine its existence. Our method does not require us to wait for the radioactive ticks of carbon-14, but measures the amount of carbon-14, directly."

The old method and the new one capitalize on the fact that all living objects incorporate a certain very small proportion of radioactive atoms of carbon-14 along with the regular, nonradioactive carbon-12.

The carbon-14 comes from certain dioxide molecules in the air made from the special form of carbon produced when cosmic rays entering the earth's atmosphere hit nitrogen atoms, converting them to carbon-14.

This process is relatively constant, and it is assumed that at any point in recent geologic time, all living things have in them the same ratio of carbon-14 atoms to carbon-12 atoms.

Once the living thing dies, however, it no longer replenishes its carbon content. The carbon-12 remains unchanged but the carbon-14, breaking down radioactivity, diminishes over time.

Because carbon-14 decays at a known rate that diminishes with its concentration, scientists can compare the rate of radioactive emissions from an ancient specimen and estimate the proportion of carbon-14 remaining. If, for example, only half the original concentration is left, the object is assumed to be about 5,730 years old because that is the time it takes half of any concentration of carbon-14 to decay.

Because the original concentration of carbon-14 is so low, about one for every trillion carbon-12 atoms, the amount left after about 40,000 years is too low to be detected by conventional methods.

Instead of counting radioactive emissions, the new method counts atoms of both forms of carbon. If the proportion of carbon-14 atoms is half that in living things, the same "half-life" figure is applied to indicate an age of 5,730 years.

In other areas of chemical analysis, atoms can be distinguished with mass spectrometers, standard devices that separate the molecules and atoms of a specimen and in effect, spray them at various spots on a surface in accordance with their masses. Such devices, however, fail to distinguish between carbon-14 and other particles that have almost the same mass, such as nitrogen-14.

Through a series of processes, including the use of a tandem Van de Graaff particle accelerator, the new method breaks down the similar molecules and separates the similar atoms. Analysts may then distinguish and compare the relative quantities of carbon-14 atoms and carbon-12 atoms to calculate the specimen's age.

*The article in the New York Times that was the first public announcement of the development of AMS and its successful application to the detection of carbon-14 in natural samples. Copyright 1977 by the New York Times Co. Reprinted by permission.*

*The Rochester MP tandem Van de Graaff control console with (left to right) D Elmore, T S Lund (chief accelerator engineer) and K H Purser.*

That same day Meyer Rubin, head of the radiocarbon laboratory of the US Geological Survey (USGS) in Reston, Virginia, phoned Rochester and talked to Purser. He said he had read the article and wondered if he could be of assistance. He said he had been waiting for years for such a development in carbon dating that could deal with small samples. As a geologist he had collected many such samples and was saving them for the day they might be measurable. In addition he had many large samples he had accurately dated by the standard decay counting method. He could provide some of them as calibrations for our new method. We eventually took him up on his offer. Rubin turned out to be a very good contact in the carbon dating field, a field in which we were neophytes.

Around the middle of June 1977 we phoned Rubin at his carbon dating laboratory to say that we thought we were in a position to actually measure some organic samples of ages ranging between modern and very old. We invited him to visit us the following Monday 20 June. He said he would bring a variety of samples that he had already dated as possible calibrations of our new technique.

Rubin and Litherland arrived in Rochester and most of the day was spent in discussions with them. One of Rubin's interests as a geologist is to study the periodicity of volcanic eruptions in the Cascade Mountains of the Pacific northwest and in Hawaii. When a volcanic mountain erupts, the lava flows over the vegetation

*M Rubin (left) and L R Kilius (right) loading a mixture of silver powder and finely ground carbon into the sputter ion source cone.*

that has grown up on that of the previous eruption. By carbon dating the organic material between lava flows one can get a measure of the periodicity of eruptions. A paper by Rubin and his colleagues appeared in a February 1975 edition of *Science* [10] reporting that such studies showed that since AD 1500 Mount St Helens had never been quiescent for more than about 150 years. Its most recent eruption was in 1857 and so the chances were good that it would erupt again, perhaps within the next few decades. On 18 May 1980 Mount St Helens blew skywards in a mighty eruption. The explosion blew off 1300 feet of the mountain's peak, killing 57 people, and devastating the surrounding 250 square miles.

Rubin brought to Rochester a number of samples he had dated of ages ranging from one he had collected at Mount Hood that was essentially modern to material collected at Hillsdale, Michigan that was 40 000 years old, about the upper limit of what he was able to date. He also told us of the work carried out a few years back by Michael Anbar [11] working at a research institute in California and funded by the National Science Foundation to develop a carbon-14 mass spectrometer system. We later found out that Anbar, employing very clever concepts including the use of negative ions, had come close to succeeding. He failed because of molecular interference. He had since abandoned his efforts.

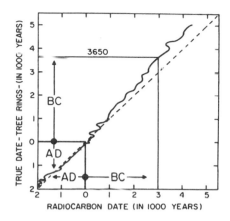

*A plot of the true date of tree rings (obtained by counting the rings) versus the radiocarbon date. This permits accurate dating back some 7000 years or more.*

Rubin mentioned many applications of carbon dating, especially if one could use small samples, in the areas of climatology, oceanography and dendrochronology. The latter is the arcane science of tree-ring dating. (See the colour section.) There is a species of tree that grows in the southwestern regions of the USA called the bristle cone pine. Some living members of this species are 4000 years old and dead logs are found whose tree ring pattern can be matched with those of living trees to provide wood going back a further 4000 years. One thus has a source of organic material that can be dated to within one year for the last 8000 years. The same samples have been used for measurements of their carbon-14 to stable carbon ratio by the decay counting method. These measurements show the cosmic ray flux incident on the earth's atmosphere is not constant but has changed somewhat with time (during the Maunder minimum it increased appreciably and, as mentioned previously, that makes it difficult, if not impossible, to date Stradivarius violins). Carbon-14 measurements on wood with accurate dendro dates provide a radiocarbon dating calibration curve, at least back 8000 years. The correction happens to be negligible for an object that is 2000 years old. However, for example, a sample that radiocarbon measurements say dates to the year 3000 BC has a true date of 3650 BC.

Litherland and I realized that Rubin with his carbon dating expertise could be very helpful to us. We phoned him inviting him to be present at our third run which was scheduled to take place starting on 4 July. He said would consult people at the US Geological Survey about appropriate carbon samples for our ion source. In particular, what would be a suitable binder to hold the carbon powder together so it would not fall out of our ion source holders? It turned out that pure, finely powdered silver worked pretty well.

Meanwhile we learned that a group at McMaster University in Hamilton, Canada, a couple of hours drive from Rochester, had started a programme to measure carbon-14 with a tandem Van de Graaff accelerator much like ours. We did not yet know when they had started, how far along they were and most importantly whether they were about to publish. We had additional inputs on Muller's progress at Berkeley and, although he had not yet measured carbon-14 in natural samples, he was a force to be reckoned with. There were no more AMS runs after our next one that could be scheduled on our machine until September. It would be very valuable to have Rubin working with us on the upcoming run.

We learned of the McMaster work in a phone call from a colleague, Parker Alford, at the University of Western Ontario in London, Ontario, Canada on 22 June 1977. He had just returned to London from an experiment at the Los Alamos National Laboratory in New Mexico. Denis Burke from the tandem group at McMaster was at Los Alamos at the same time and Alford had shown him the article on our work in the *New York Times*. He told Alford that he and a former student of his, Erle Nelson, had used the McMaster tandem, a smaller version of our accelerator, to measure carbon-14 in a sample of nineteenth century wood.

That same day I received a phone call from Litherland who was in Saskatoon at a meeting of the Canadian Association of Physicists. He had just heard about the Burke–Nelson work at McMaster from the director of the McMaster tandem laboratory. Litherland seemed pretty sure that the inspiration for the McMaster work had originated somewhere else and that, as we later learned, was correct.

Shortly after talking to Litherland I called Burke at Los Alamos and he confirmed that they had had a run at virtually the same time as ours. It later turned out to have taken place two weeks after our first successful run. He said that Nelson, after getting his PhD in nuclear physics under Burke on the McMaster tandem, had been appointed as an assistant professor in the Department of Archaeology at Simon Fraser University in British Columbia, Canada. It was Nelson who had suggested the experiment to

Burke and Burke had merely given him a hand in carrying it out. He said he thought Nelson was preparing a paper on the work to be submitted to *Science*. We later learned that Nelson mailed his paper to *Science* on 22 June, the same day I had talked to Alford and Burke.

We finally got our article to *Science* completed and we mailed it to them on 24 June. That day we were contacted by two members of a group at the University of Arizona that would play a very important role in AMS carbon dating in future years. They were Paul Damon and Austin Long who were members of the Laboratory of Isotope Geochemistry in the Department of Geosciences at Arizona. They had read of our work in the *New York Times* and wanted more information.

For many years that university had been a leading centre for carbon dating and in the study of dendrochronology. Damon and Long turned out to be two top scientists in the conventional carbon dating field. With our support and urging, the University of Arizona would have the first tandem accelerator mass spectrometer in the world dedicated to carbon dating, an instrument that would be built by Purser's company.

That same day the issue of *Time* magazine dated 27 June 1977 containing the article on the new carbon dating method appeared on the news stands. It had been written by Peter Stoler who had phoned us for an interview shortly after the article had appeared in the *New York Times*. The article was entitled, in the usual brash *Time* style, 'The new dating game'.

On 28 June Ted Litherland told us that, as a potential user of the McMaster tandem facility, he regularly received a schedule of the experiments to be carried out there. The one he had received in May showed time allocated to Burke and Nelson for 'carbon beam tests' for the dates of 1 and 2 June. There was no indication that this meant carbon-14 and so Ted had no advance warning of the McMaster competition with us. Erle Nelson's inspiration, he freely acknowledged, came from his contacts with Rich Muller and any knowledge he may have had of our work could only have had the effect of hastening the submission of his McMaster results to *Science*.

We phoned our contact editor J E Ringle at *Science* on 29 June to see whether our manuscript had arrived. It had not, which was surprising since it had been mailed five days earlier. At that time the US mail was rather slow and the Canadian even slower, or so we hoped.

## New Dating Game

Archaeologists and paleontologists trying to ascertain the age of bone, wood and charcoal from ancient sites have long employed a technique called carbon-14 dating. This dating game has its drawbacks: it requires the destruction of a sizable portion of the sample and cannot, without costly and time-consuming treatment, determine the age of any object more than about 40,000 years old. But a new method promises to overcome both obstacles. A team of researchers from the University of Rochester, the University of Toronto and General Ionex Corp. of Ipswich, Mass., is developing a way of dating objects that not only uses much smaller samples, but may also more than double the age that can be evaluated.

The method does not do away with the need to measure carbon 14, a radioactive atom that accumulates in all organisms while they live and decays at a known rate once they die. But it measures it in a different way. Current dating methods determine the age of an object indirectly, by measuring its carbon-14 radioactivity. The new technique

being developed by Professor Harry Gove of Rochester and his fellow researchers measures the amount of carbon 14 directly. The scientists place a sample of the object to be evaluated in Rochester's tandem Van de Graaff particle accelerator. The machine separates carbon 14 and carbon 12, an atom that also accumulates at a steady rate but does not decay, from all other atoms in the sample. By comparing the ratio of these two types of carbon, the researchers can then calculate the age of the object under study.

**Tiny Samples.** Gove believes the direct measurement system, which requires as little as one-hundredth of the material needed for current dating tests, will eventually win wide acceptance. He and his colleagues have accurately determined some test samples to be 70,000 years old. With more work, they believe, they can push radiocarbon dating of tiny samples back to 100,000 years.

TIME, JUNE 27, 1977

*An article on the May 1977 Rochester AMS work published by Time magazine and written by Peter Stoler. Copyright 1977 by Time Inc. Reprinted by permission.*

Ringle suggested we call him the next day which we did and he said our paper had arrived in the morning's mail. However, he noted that a paper by Nelson, Kortelling and Stott on the same subject had been received the day before. In their cover letter to *Science* they said they had carried out the experiments on 1 and 2 June 1977 which we had already surmised and that they had learned of our work from the *New York Times*. Ringle said the two papers would probably be published in the same issue of *Science* with the Nelson paper paginated before ours since it was received a day earlier—and that is what happened.

The McMaster paper entitled 'Carbon-14: direct detection at natural concentrations' appeared on p 507 [12] and ours titled 'Radiocarbon dating using electrostatic accelerators: negative ions provide the key' on p 508 [13] of the 4 November 1977 issue of *Science*. One difference between the two papers was the lack of

*AMS data display. The three peaks left to right are carbon-12, carbon-13 and carbon-14. The people are H E Gove (left) and D Elmore (right).*

a background measurement in the McMaster work. Their paper suggested that the carbon-14 they detected was coming from the nineteenth century wood sample they used but did not prove it. On the other hand the Rochester tandem experiment produced the first ever carbon-14 date by accelerator mass spectrometry, since we made a background measurement on graphite. The work at Rochester had been carried out two weeks before that at McMaster but they beat us by one day in getting their manuscript to the publisher.

Why this concern about publication priority in a journal like *Science?* A paper on our work had already been read at an international conference and would, in due course, be published in the proceedings of that conference to prove it. One of the three senior scientists on our project (Purser) had a patent application which essentially outlined the principles of the technique. Was that not enough to establish priority for us? Maybe it was. On the other hand the patent made no mention of the application of tandem accelerators to the detection of carbon-14 or to any other radioisotope. As far as publication in a scientific journal is concerned it helps if the articles are refereed by outside experts in the field. Some conference proceedings do have their papers refereed but, so far as we knew, the Strasbourg meeting did not. Although, in fact, the Rochester–Toronto–General Ionex consortium had been the first to measure carbon-14 in natural samples, technically Nelson could claim priority. In the end it did not matter very much.

In a period of a little over six weeks a revolutionary new method of carbon dating had been developed to the point of producing the first carbon date by AMS. The results had been presented at an international scientific meeting and submitted to a scientific journal. Widespread publicity in the world press had ensured that the technique came to the attention of the world's carbon dating community. Both they and their users began to respond enthusiastically. We were now getting a flood of requests, many from scientists, to date all sorts of weird and wonderful objects far too small for the conventional method to handle. To all of these we replied with a standard letter that we were not ready yet.

I received an invitation to talk on our AMS work at the American Physical Society's nuclear physics divisional meeting in October that was to be held at the University of Rochester. As organizers we could choose a banquet speaker. I had thought of the University of Arizona's Paul Damon as a possibility. He had been one of the first carbon daters to contact us and when I suggested him Purser and Litherland were enthusiastic. I called Damon and he agreed to talk on 'Radiocarbon dating and the unity of science'.

We received a letter dated 23 November 1977 from Professor Stanley N Davis of the University of Arizona's Department of Hydrology and Water Resources inviting us to attend a workshop on 16–18 March 1978 at Tucson on methods of dating old ground water. This is a scientific area of considerable importance impacting on questions of water flow rates in aquifers and on the integrity of potential nuclear and chemical waste disposal sites. It was this workshop that got us launched into what became the most important AMS effort at Rochester, namely the detection of the cosmogenic radioisotope chlorine-36.

Rubin was present during our third AMS run in July 1977 and we measured several samples he had previously dated. In late November and early December we completed the measurements on Rubin's samples and these constituted data for our second paper to *Science*. It was entitled 'Radiocarbon dating using electrostatic accelerators: dating of milligram samples' [14]. It was hand delivered at the end of the run to the main Rochester post office and sent by express mail to its editor J E Ringle with whom we had dealt in our first submission to *Science*.

**Table 2.1.** *The first demonstration of the dating of milligram samples by AMS. The samples weighed 15 milligrams except for the Mount Hood sample which weighed 3.5 milligrams [14].*

| | Age in years | |
|---|---|---|
| Sample | Rochester | USGS |
| Mount Hood | $220 \pm 300$ | $220 \pm 150$ |
| Mount Shasta | $5700 \pm 400$ | $4590 \pm 250$ |
| Lake Agassiz | $8800 \pm 600$ | $9150 \pm 300$ |
| Hillsdale | $4100 \pm 1000$ | $3940 \pm 1000$ |
| Graphite | $48\,000 \pm 1300$ | |

On 8 December we were informed in a phone call from *Science* that our manuscript had been received. This paper [14] would be the first to present results of AMS carbon dating of organic samples of ages ranging from modern to 40 000 years, again with a background measurement on graphite. The sample sizes employed ranged from 3.5 to 15 milligrams. The agreement between our dates and Rubin's was excellent and the accuracy was comparable. Joint press releases on the work were issued by the Universities of Rochester and Toronto and good coverage ensued in both US and Canadian newspapers.

# Chapter 3

# Historical Development of Accelerator Mass Spectrometry—1978–80

We now turned to the task of organizing the first international AMS meeting which we planned to hold in Rochester on 20 and 21 April 1978. Purser, Litherland and I would be the co-chairs. At my suggestion the meeting was to be called 'The First Conference on Radiocarbon Dating with Accelerators' [1]. It was not a felicitous choice. In the event, papers were presented on the detection by AMS of two other cosmogenic radioisotopes in addition to carbon-14. These were beryllium-10, by a group from the René Bernas Laboratory in France, and chlorine-36, by the Rochester group. Purser made this point when he agreed to be a co-chair as did Muller at Berkeley when we phoned him inviting him to be on the organizing committee. Subsequent meetings in this field are now called international conferences on accelerator mass spectrometry.

In the call to Muller he said they were still having background problems. He mentioned that on 6 December 1977 they made a blind measurement of a sample they had obtained from Rainer Berger at UCLA (University of California, Los Angeles). They would be sending their results to Berger to see how close they had come. He said he would be submitting a paper on the results to *Science*. Making blind measurements at this stage, especially on the Berkeley cyclotron, was risky because of the possibility of getting it wrong. If that happened it might tend to discredit AMS.

During late January 1978 the Rochester consortium showed that the mass 14 molecule carbon dihydride ($^{12}CH_2$) was sufficiently stable when two electrons were removed from the neutral

24

molecule to cause it to interfere with the detection of carbon-14 [2, 3]. That meant Purser's tandetrons might have to be designed with terminal voltages of 2 MV to remove three electrons and, thus, be available at a concomitantly higher price. However, as mentioned in Chapter 11, a way of circumventing this was developed in 1998.

In mid March the three day workshop on the dating of old ground water mentioned in Chapter 2 was held at the University of Arizona. It was attended by some 34 people including Rubin, Purser and myself. The Berkeley effort was represented by T S Mast, one of the Muller collaborators. The workshop participants concluded that the development of a method to directly detect chlorine-36 and the continued development of the carbon-14 method held the most promise for dating old ground water. The proceedings were published by the University of Arizona [4].

At Rochester we had already begun to think about how we should proceed to measure chlorine-36. Stable isobars were the main problem. Isobars are different chemical elements with the same mass. Unlike carbon-14 which has no interfering stable isobar because negative nitrogen-14 ions are unstable, chlorine-36 has a stable isobar, sulphur-36. Its negative ions are as stable as those of chlorine. The other stable isobar argon-36, because it is a noble gas, does not form negative ions. Sulphur is a very common element and it would be a major problem to detect chlorine-36 at the exquisitely low concentrations required in the presence of a flood of sulphur-36. It is a problem that was eventually solved.

From 10–18 April 1978, inspired by the conclusions reached at the Arizona meeting, we tried to measure chlorine-36 in a natural sample, namely salt from the flats at the Great Salt Lake in Utah. In addition we had some carbon tetrachloride that had been enriched in a reactor to have a chlorine-36 to stable chlorine ratio in the range of one part in a trillion. We were readily able to detect the chlorine-36 in the latter but not in the natural salt sample. However, we obtained enough information on how to improve our apparatus to make the measurements with sensitivities increased by two or three orders of magnitude. We made another unsuccessful attempt in mid May and finally achieved success in late July. It should be noted that chlorine-36 is the only important cosmogenic radioisotope that requires for detection a tandem accelerator with terminal voltages approaching 10 MV. The reason for this will be explained in Chapter 5.

A couple of days before our April carbon dating conference was to start I met Christine Waterhouse, a professor of medicine, and Julian Kielson, a professor of statistics, both at the University of Rochester. Waterhouse had been conducting some studies on the metabolism in human subjects of glucose tagged with carbon-14. Because she was using the standard decay counting method to measure the radiocarbon in blood samples taken from the subjects, rather high doses of carbon-14 and large blood samples were required. Our new detection method should be able to substantially reduce both. Kielson had been assisting Waterhouse in the analysis of her data. I suggested this would make an interesting contribution to our conference and Kielson agreed to give a talk on the subject. It would be the first suggestion of applications of AMS in medical research. Years later this would become a very important research area for AMS. This is discussed further in a later chapter.

There were 84 delegates in attendance at this first AMS international meeting of whom 19 came from various European countries including France, West Germany, the Netherlands, Sweden, Switzerland, and the United Kingdom, 15 from Canada and the remaining 50 from the USA. Some of the more notable delegates were Rainer Berger, the UCLA carbon dater who supplied Muller with blind samples, Jurg Beer from Willem Mook's carbon dating laboratory in Groningen, one of the world's best, Minze Stuiver, also from Seattle, from another of the top carbon dating laboratories, E T Hall, the Oxford entrepreneur with his colleague Robert Hedges who would eventually run Hall's AMS laboratory in Oxford, Lucy McCrone, the wife of Walter McCrone who claimed the Vinland map was a fake, Richard Muller, Erle Nelson, Grant Raisbeck who pioneered measurements on beryllium-10, and William Rodney of the National Science Foundation who provided support for the AMS effort at Rochester. There were reports on the detection of various cosmogenic radioisotopes, mostly carbon-14, but also beryllium-10 and chlorine-36. The proceedings [1] were published remarkably quickly in Rochester.

It was the first in a series of meetings now called international conferences on accelerator mass spectrometry and held every three years in the spring. The second [5] in the series was held at Argonne National Laboratory near Chicago in 1981 and the third [6] in Bern–Zurich in Switzerland in 1984. The fourth [7], called Anno Decimo, was held in 1987 at Niagara-on-the-Lake in Ontario,

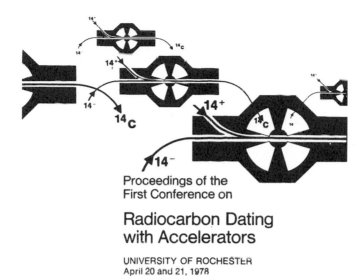

Proceedings of the
First Conference on

## Radiocarbon Dating
## with Accelerators

UNIVERSITY OF ROCHESTER
April 20 and 21, 1978

Edited by H. E. Gove

*Cover page of the Proceedings of the First Conference on Radiocarbon Dating with Accelerators. These are now called international conferences on accelerator mass spectrometry.*

Canada more or less geographically midway between Hamilton (McMaster), Toronto and Rochester. It was organized jointly by AMS people from those three institutions. It celebrated the tenth anniversary of the first successes of AMS. The fifth [8] was held in Paris in the spring of 1990, the sixth [9] in Australia in the fall of 1993 and the seventh [10] in Tucson, Arizona in May, 1996. The eighth will be held in Vienna, Austria in 1999. Since 1954 international radiocarbon conferences have been held in various parts of the world approximately every three years. Since the invention of AMS, starting with number 10 in Bern, Switzerland and Heidelberg, Germany, they have devoted a fair fraction of their meeting time to that field of AMS. Number 15 in this series was held in Glasgow, Scotland in 1994. The latest one, number 16 took place in 1997 in Groningen, the Netherlands.

It was during our chlorine-36 run in May 1978 that Litherland, Purser and I agreed that we should explore the possibility of getting involved in dating the Turin Shroud. We had learned of it in June 1977 in a letter from a member of the British Turin Shroud Society. I had already privately decided that it would be too good an opportunity to miss. It would be a highly public demonstration of the power of carbon dating by AMS. Besides, I was becoming increasingly curious about the shroud and the, to me, remote possibility that it actually was Christ's burial cloth. I had not the slightest inkling how byzantine the project would turn out to be nor that it would be 10 years almost to the day that we agreed to be involved in the Turin Shroud adventure before it was first dated by AMS at Arizona. The complexities of this enterprise are recounted in the book *Relic, Icon or Hoax? Carbon Dating the Turin Shroud* (see reference [1] in Chapter 1). A summary is to be found in a later chapter.

Meanwhile we had begun to wonder about the second paper we had mailed to *Science* on 5 December 1977. A *Science* editor had informed us that it had been received on 8 December. We had learned in January 1978 that one of the referees of the paper was Meyer Rubin. When I phoned the *Science* editor J E Ringle in early February I was told that he had received the report from one referee who liked it a lot and that he would jog the other referee. I learned from Professor Minze Stuiver at our Rochester AMS conference in April that he was the second referee. Stuiver is the head of the crack conventional carbon dating laboratory in Seattle. We had already responded on 24 March to some useful comments he had made concerning our manuscript. He said he had informed *Science* that he was now happy with the paper.

By now it was mid May, and over six months since *Science* had received our paper. We phoned Ringle regarding the status of the paper but he was unable to locate it. He said he would have someone call me back. Their manuscript editor Lois Schmidt did call back to say that the manuscript had been misplaced but that she had somehow found it.

The inordinate delay in dealing with our paper by *Science* was not because they had misplaced it. Major journals like *Science* do not misplace manuscripts. The real reason was a letter they had received from a member of an AMS group to which we had provided a preprint. In the letter the writer objected to the way in which his group's contributions had been referenced. Schmidt gave us the opportunity of responding to the complaint. We declined to do so on the grounds that our references had been both appropriate and fair. She was very apologetic and said she would edit it immediately and send it to the printer. We should get proofs in five or six weeks.

Litherland, Purser and I also discussed where we should publish our chlorine-36 results when we finally got some decent ones. This, we were sure, would be another paper that would

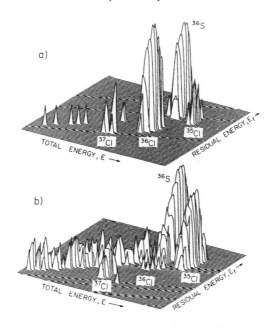

*A two parameter logarithmic spectrum for AMS measurements of chlorine-36 in (a) a sample of water from Lake Ontario and (b) zone refined reagent grade AgCl [13].*

merit publication in a prestigious journal. We excluded *Science* as a possibility because of the long delay in publishing our second paper. The only other possibilities we considered were *Physical Review Letters*, which is considered by researchers in the physical sciences to be the place to publish, and the British journal *Nature*. We decided to try *Physical Review Letters* first. In early May I called the senior editor of *Physical Review Letters*, George Trigg, whom I had got to know in my capacity as associate editor for nuclear physics of that journal, a position I was still filling. *Physical Review Letters* has a category on geophysics which I thought might be appropriate and he agreed.

The experiments that finally produced the data for this chlorine-36 paper took place on 20–23 July 1978. A paper on the results was submitted to Trigg. The ratio of chlorine-36 to stable chlorine was measured in water samples from various sites in Arizona and from the Lake Ontario water intake for Eastman Kodak. We had solved the problem of separating the cosmogenic radioisotope chlorine-36 from its stable isobar sulphur-36 by a combination of chemical and nuclear techniques. We became the first group to detect this important cosmogenic element by AMS, again using the extraordinarily small sample sizes characteristic of AMS. The demand for such measurements is now considerable. For example, as recounted in a later chapter, it was used to measure the slow neutron doses received by the victims of the 1945 atomic bombing of Hiroshima.

Around this time we also sent a paper to Turin describing how we would go about dating the shroud and offering to do so if the authorities there wished. This paper was published in the proceedings of an international congress on the Turin Shroud held in Turin in October 1978 [11].

From 9–13 September 1978 we had another AMS run. This time we measured carbon-14 in a small piece of linen that Meyer Rubin had obtained from the Smithsonian. It was taken from the cloth wrapping of an Egyptian bull mummy and was known to date from around AD 0. He had dated it to be $2050 \pm 200$ years old. Our result was $2200 \pm 150$ years [12]. Clearly we would have no problem in dating the Turin Shroud even if it were 2000 years old, although it would be useful to pin its age down more accurately.

Eventually *Physical Review Letters* rejected our chlorine-36 paper. So much for my influence as one of the associate editors of the journal! It was submitted to the British journal *Nature*. They accepted the paper, entitled 'Analysis of chlorine-36 in environmental water samples using an electrostatic accelerator', on 4 October 1978 and it was published on 4 January 1979 [13]. Like our first paper on carbon-14 it was a seminal work that would have great influence on the AMS field in the years to come.

From 20–26 October 1978 we had another AMS run in which we measured another cosmogenic radioisotope, beryllium-10 [14]. It was during this run that James Arnold of the Chemistry Department of the University of California at San Diego phoned. Arnold had worked with Willard Libby in the early days of radiocarbon dating in the mid 1940s and was now a leading figure in the field of cosmogenic isotopes. He had been one of the speakers at the first AMS conference in Rochester in April 1978. He had read the preprint of the chlorine-36 paper that had been rejected by *Physical Review Letters*. He was very impressed with it and wanted to start a collaborative programme of research involving the measurement of this radioisotope in Antarctic meteorites. Arnold is one of the pioneers in the measurement of radioisotopes in extraterrestrial material. (See the colour section.) His expressed interest in our chlorine-36 AMS work strengthened our feeling that it was a significant piece of research and that

*Physical Review Letters* had erred in rejecting it. However, to have it published in *Nature* was even better.

We had another AMS run from 29 January to 1 February 1979. It was on chlorine-36 and included samples of Antarctic meteorites and Antarctic ice from Jim Arnold of the University of California at San Diego [15,16] and some important underground water samples from Harold Bentley of the University of Arizona.

In early February 1979 Lloyd Currie of the National Bureau of Standards (now called the National Institute of Science and Technology) and I discussed how we might go about measuring the carbon-14 to stable carbon ratio in air samples of both methane and carbon dioxide that he had collected. These would have carbon contents a hundred times smaller than anything we had run so far, i.e. a few tens of micrograms of carbon instead of a few milligrams. Such small samples would only last for a few minutes in our ion source before they were exhausted, but this might be enough time to obtain the accuracy he required. We had already achieved a factor of a thousand reduction in sample size over the conventional Libby decay counting method and when we ran his samples a year later we achieved a factor of a hundred further reduction. We achieved a 3% accuracy for these samples that had a carbon-14 to carbon ratio of about one part in a trillion [17].

Some time in February, Morris Goodman, a professor of anatomy at Wayne State University in Detroit, Michigan, phoned me concerning a 350 milligram sample of muscle from a baby woolly mammoth he had received from some colleagues of his in the USSR. He wondered if we would be willing to carbon date it and how much sample we would require. This creature, dubbed Dima, had been discovered by a bulldozer operator in a peat bog in the Siberian permafrost in 1977 and had been recovered whole from the site. It had become trapped in the bog and had frozen to death. It was almost perfectly preserved, as a photograph he later sent me vividly demonstrated. Goodman had found whole red and white blood cells in the sample and he was hoping to shed light on the question of whether the woolly mammoth was the ancestor of either the African or the Indian elephant. I told him we would indeed be interested in dating it but that he should call Rubin to see how much sample we would require. It turned out that 50 milligrams was an adequate amount.

In mid February Arnold called. He said that a ship from Antarctica would be docking soon in San Francisco with five meteorites as well as samples of the ice on which they were found.

*Dima, a baby woolly mammoth discovered in 1977 in a peat bog in the Siberian permafrost. It had become trapped in the bog and had frozen to death. It was almost perfectly preserved.*

He would like us to measure chlorine-36 and beryllium-10 in both the ice and the meteorites. These measurements could tell us how long ago the meteorites had fallen onto the ice and also how old the ice was.

On 20 February 1979 Rubin arrived for an AMS run. He had converted 63 milligrams of baby woolly mammoth muscle to about 15 milligrams of carbon. He said that, in the process, a most obnoxious odour permeated his laboratory causing one of his colleagues to announce that a mammoth barbecue was under way. Meyer had also brought some more Egyptian bull mummy shroud samples. We made a preliminary measurement of the mammoth muscle and found it to be greater than 23 000 years old [12]. We were having some background problems so this age was a lower limit. I passed the information on to Goodman.

On 22 February 1979 Donald Hess, the University of Rochester's Vice President for Campus Affairs, sent me a copy of the testimony given by James Krumhansl, a senior official of the National Science Foundation, before the US House Committee on Science, Research and Technology on 14 February 1979. Hess's note read in part 'Jim Krumhansl...used an example of your work to illustrate some very good science. Few universities were mentioned in Krumhansl's testimony; you should feel honored!'

**Table 3.1.** *Recent (1979–80) Rochester AMS results [12]. The woolly mammoth Dima was older than 23 000 years. The date for the bull mummy cloth indicated that a date for the Turin Shroud was possible, but better accuracy, which was achieved, would be needed.*

| Sample | Rochester age (years BP)[a] | Expected age |
|---|---|---|
| Mount Shasta | 4580 ± 90 | 4590 ± 250 [b] |
| Bull mummy cloth Dashur, Egypt | 2200 ± 150 | 2050 ± 200 [b] |
| Baby woolly mammoth (USSR) | > 23 000 [c] | Available sample size too small to date by conventional methods |
| Australian sucrose | 1.47 ± 0.03 [d] | 1.45–1.51 [e] |

[a] BP = years before 1950. $^{14}C$ half-life assumed to be 5568 years.
[b] As measured by the USGS.
[c] Preliminary result.
[d] This is the ratio of $^{14}C/^{12}C$ in the sucrose standard to the $^{14}C/^{12}C$ ratio in 1950 wood.
[e] Measurements of this ratio from other laboratories fall in this range.

In his testimony Krumhansl mentioned advances in two areas of physics. The first was the development of the ion trap technique which made it possible to isolate and store individual ions and electrons in a vacuum for periods of several days so that precise measurements of their properties could be made. Ten years later, in 1989, this earned the co-inventors of the technique, Hans Dehmelt of the University of Washington and Wolfgang Paul of the University of Bonn, the Nobel Prize for physics.

Krumhansl's statement then went on to read 'Secondly, major advances in particle and nuclear science have also occurred. One of these has aroused intense interest because of its broad range of applications in other fields. The work, which was a cooperative effort between the University of Rochester, the University of Toronto, and the General Ionex Corporation showed that a tandem Van de Graaff accelerator, normally used for nuclear physics research, can be used as an extremely powerful ultrasensitive detector of minute traces of chemical elements. The immediate application has been to extend the sensitivity of carbon-14 dating to about 70 000 years using samples one-thousandth the size of those previously used. Other applications envisaged range from studies of ground water movements to vast improvements in medical diagnostics.' We were very pleased to have received this

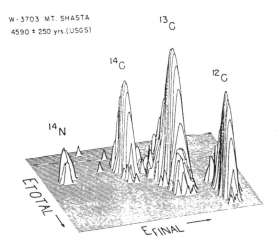

W - 3703 MT. SHASTA
4590 ± 250 yrs.(USGS)

$^{13}C$

$^{14}C$

$^{12}C$

$^{14}N$

$E_{TOTAL}$

$E_{FINAL}$

*A two-parameter logarithmic plot of the carbon-14 measurement of an organic sample from Mount Shasta. The Rochester result was a substantial improvement over the result referred to in Chapter 2 and was even better than the USGS decay counting result despite using a sample size about 1000 times smaller.*

recognition and, in retrospect, to have it coupled with an advance in physics that later won the Nobel Prize was high praise indeed.

The Tenth International Radiocarbon Conference [18] was held from 19–26 August 1979 with the first half taking place in Bern, Switzerland and the second in Heidelberg, Germany some 300 km away. These meetings have been held approximately every three years since Willard Libby invented radiocarbon dating in the mid 1940s. This, however, was the first of them to take place since the invention of accelerator mass spectrometry and its application to carbon dating in May 1977, and there was one session at the Heidelberg part of the meeting devoted to this new technique that was to completely revolutionize the field.

There were seven AMS papers in this session. I gave the lead-off paper which was co-authored by the others in the Rochester/Toronto/GIC group and Meyer Rubin of the US Geological Survey. I discussed the work we had carried out in the past two years on the detection of carbon-14, chlorine-36, beryllium-10 and aluminium-26. There was one by Ken Purser on his plans for manufacturing a small tandem of 2.5–3 MV terminal voltage to be supplied to Arizona, Toronto and Oxford and to be used mainly for carbon dating, ones from groups at Argonne National Laboratory, the Chalk River Laboratory of Atomic Energy of Canada, the UK Atomic Energy Research

Establishment at Harwell and the University of Washington in Seattle all describing measurements they had made on carbon-14 and in the case of Argonne also on aluminium-26, silicon-32 and chlorine-36 and a paper by the Oxford group describing the plans for their facility based on a Purser accelerator.

On 12 October I visited Jim Arnold and his colleagues at the University of California in San Diego. Before I left, Arnold told me he was having some iodine-129 samples from meteorites prepared for us to measure.

Iodine-129 is another radioactive element produced by cosmic ray interactions, in this case with xenon, in the atmosphere and by certain cosmic ray interactions with elements in meteorites. It has a half-life of 15.7 million years. Among other applications it has the potential for tracing and dating oil deposits. It is also a long-lived fission product produced in nuclear reactors and is released to the environment in nuclear reactor accidents. Its only stable isobar is xenon-129 which does not form a stable negative ion and that fact makes it an obvious isotope for detection by AMS.

Between 22 and 28 October 1979 we measured $^{129}I/I$ ratios in artificial samples and also in meteorite samples prepared by Arnold's group and water samples prepared by Harold Bentley of Arizona. This was another first [19]. Our best sensitivity at the time seemed to be a few parts in $10^{13}$. During this run we also carried out the first application of AMS to the detection of low concentrations of stable elements. We measured stable platinum isotopes in two rock samples with known concentrations of platinum of 4 and 0.06 parts per million and verified the appropriate ratio of concentrations [20]. We also switched between the two isotopes platinum-194 and platinum-195 and got the correct one to one ratio of the two that occurs in nature [21].

During another visit Litherland made to Rochester, we discussed the iridium measurements that Luis Alvarez and his colleagues had made, using a technique called neutron activation, in a layer of sediment that dated to the Cretaceous–Tertiary (K–T) boundary 60–65 million years ago [22]. Neutron activation is a procedure in which the iridium-containing material is inserted in a reactor. The iridium captures neutrons to form easily identified radioactivity. This K–T boundary layer is well defined geologically in many parts of the world. Iridium is an element that has a low abundance in the earth's crust but is much more plentiful in meteors and other extraterrestrial material. Alvarez concluded

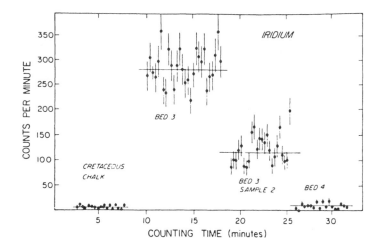

*Rochester AMS measurements of iridium-193 in sediment samples from Stevns Klint, Denmark. The samples were taken in ascending order from the Cretaceous chalk layer 2.2 m below the surface through the Bed 3 sample within the Cretaceous–Tertiary boundary layer and the Bed 3 sample 2 and Bed 4 material from layers immediately higher up.*

that a giant meteorite had impacted the earth at that time and had caused a cloud of dust to rise in the air and to surround the earth. This cloud blocked the sun's rays causing plants and the dinosaurs that fed on them to die. The dust, enriched in iridium, settled to earth and became part of the K–T boundary. This cause of the dinosaurs' demise, however, has not received universal acceptance [23].

Platinum is also an element that is depleted in the earth's crust but is more abundant in meteorites for the same reason as iridium. Therefore any iridium-enriched sediment should also be enriched in platinum. We had already made preliminary measurements of platinum in other samples as mentioned above and we later measured platinum and iridium in the sediment used by Alvarez and confirmed both were highly enriched within the K–T layer [21, 24, 25].

Sometime in the summer of 1980 using a combination of AMS and decay counting we made a measurement of the half-life of silicon-32, a cosmogenic radioisotope of interest to oceanographers [26]. On 24 June Litherland called me from the Brookhaven National Laboratory. He said that a similar measurement by a Brookhaven/Argonne group had been submitted to *Physical Review Letters* on 16 May; that their value was $101 \pm 18$ years [27].

Our number, $108 \pm 18$ years, was in good agreement within the uncertainties. The two papers appeared simultaneously in *Physical Review Letters*.

Erle Nelson, who headed the Simon Fraser University AMS consortium at McMaster, wrote to me sometime in July 1980. He enclosed a preprint of an injection system that they had designed for their carbon-14 and beryllium-10 work. He wanted us to make some technical comments on it. He ended by saying that he had just received a preprint of our work on the silicon-32 half-life. The last value for the half-life he had heard was 650 years and that our new measurement of around a hundred years should really put 'the cat in the pigeon coop' for a few oceanographers. I wrote back saying that we were pleased that the same result had been obtained quite independently by the Argonne–Brookhaven group.

Litherland, Purser and I received a letter dated 13 August 1980 from Pergamon Press, whose headquarters are in Oxford, England. It was signed by Robert Maxwell, the publisher and Dr H Seligman, the editor. It stated that the Board of Editors of the *International Journal of Applied Radiation and Isotopes* (JARI) voted to give the first JARI award (a suitably inscribed silver medal and a parchment scroll) jointly to Dr K H Purser, Professor A E Litherland and me. A copy of the citation was enclosed.

The citation read: 'The recipients are honored for their contributions to the development of accelerator based dating techniques that allow the direct detection of long-lived isotopes rather than waiting to observe their decay as in previous methods. In particular for carbon-14, they carried out the key experiment demonstrating that negative nitrogen ions are too fragile to survive passage through tandem electrostatic accelerators. This allows the direct detection of carbon-14 without interference from a much larger amount of nitrogen-14 that is always present. Direct detection has made possible radiocarbon dating of samples orders of magnitude smaller than needed for beta decay measurements and in much shorter counting times. Subsequently the recipients and others have shown that the method can be extended to other cosmogenic radio nuclides including beryllium-10, chlorine-36 and aluminum-26, opening the way to new developments in archaeology, geochronology and climatology. The wide spread acceptance of accelerator based isotope dating testifies to its importance. Many laboratories are now active in the field and several dedicated facilities are under construction.' The publications cited were [3], [9] and [12] listed in Chapter 2.

*Presentation of the JARI award in April 1981. From left to right: H E Gove,*
*an official of the Journal of Applied Radiation and Isotopes, K H Purser and*
*A E Litherland.*

The award was presented to us at the Third International
Conference on Electrostatic Accelerator Technology held in Oak
Ridge, Tennessee on 15 April 1981.

The use of a positive ion cyclotron for AMS, such as the
one at Berkeley (reference [4], Chapter 2), received the *coup de
grâce* at the hands of its proponents when they attempted to
make AMS measurements using blind samples. A blind sample
in this case is one that has been measured at a conventional
carbon dating laboratory. It is supplied for measurement with
no information provided on its age. The result is then sent back
to the laboratory that supplied it, for verification. The first blind
sample, mentioned previously in this chapter and supplied by
Rainer Berger, had been measured by him using the method
invented by Libby (reference [2], Chapter 1). Berger's value
was 5080 ± 60 years. The Berkeley group obtained a value of
5900 ± 800 years [28]. The second blind sample measured by
Berger to have an age of 18 800 ± 2500 years was measured
by Berkeley to be 8000 ± 600 years old [29]. In both cases,
remarkably enough, Muller mailed his results to me at the
same time as he sent them to Berger. These results did not
have a deleterious impact on the science of AMS but merely

demonstrated the unsuitability of positive ion cyclotrons for making AMS measurements.

In the 48 months since the first AMS work in May 1977 at Rochester by the General Ionex/Rochester/Toronto consortium that group made notable contributions to the AMS field. These included the first measurement of carbon-14 in a natural organic sample, the dating of organic samples back to 40 000 years, the first measurements of chlorine-36 in natural samples, the first measurements of iodine-129 in natural samples, the determination that carbon dihydride lived too long when only two electrons were removed to prevent tandem carbon dating accelerators with terminal voltages lower than 2 MV to work efficiently, the holding of the first international AMS conference, the offer to date the Turin Shroud, the measurement of chlorine-36 in Antarctic meteorites and ice, the measurement of carbon-14 in atmospheric methane samples as small as 50 micrograms, the citation in Congressional testimony for our AMS work, the first AMS measurements of stable isotopes in the Cretaceous–Tertiary boundary, the first half-life measurement of a cosmogenic radioisotope (silicon-32) by AMS and the award of the Pergamon Press JARI medal.

At the first AMS conference in April 1978 [1], in addition to the three laboratories at Berkeley, McMaster and Rochester, results of AMS measurements were presented by scientists from the Rene Bernas Laboratory in Orsay, Atomic Energy of Canada Ltd at Chalk River and Oxford University. By early 1981, at the time of the second AMS conference [5], several other nuclear physics accelerator laboratories had joined those mentioned above in devoting some of their accelerator time to AMS measurements. These included the Argonne National Laboratory, the University of Pennsylvania, the University of Washington, the Weizmann Institute of Science in Rehovot, Rutgers University and the Laboratory for Nuclear Physics in Zurich. Tandem accelerator facilities specifically dedicated to AMS measurements were well along in the planning stages at the University of Toronto, Oxford University, the University of Arizona and Nagoya University.

It had been an exciting four years.

# Chapter 4

# The Development of Tandem Electrostatic Accelerators

Accelerator mass spectrometry was developed by nuclear physicists and generally employs techniques that are familiar to those in the field who use beams of charged particles from accelerators in their research. This chapter and the following are written not for such people but for those in research areas such as archaeology, geology and other non-nuclear fields who employ AMS as a measurement technique and who would like to have a better understanding of AMS, of its potential and of its limitations.

By the early part of 1997 there were 40 AMS facilities operating throughout the world, one in Canada, eight in the USA, fifteen in Europe, ten in Asia, four in Australia and New Zealand and two in South America. Of these all but two (one in the USA and one in China) are based on tandem electrostatic accelerators. Clearly tandems have become the accelerator of choice for AMS.

Tandem accelerators have been employed in basic nuclear physics research since October 1958 when the first tandem Van de Graaff machine underwent factory tests of its operating characteristics at the High Voltage Engineering Corporation (HVEC) in Burlington, Massachusetts. During these tests the first nuclear physics research employing a tandem electrostatic accelerator was carried out [1]. This accelerator, designated by HVEC as an EN tandem, was built for Atomic Energy of Canada Ltd (AECL) at Chalk River, Ontario, and was installed there as a nuclear physics research instrument in February 1959. At about the same time two 5 MV tandems with similar characteristics to the HVEC EN tandem were built by Metropolitan-Vickers Electrical Co., Ltd and installed at the United Kingdom Atomic Energy

Authority's laboratories at Harwell and Aldermaston. The Chalk River EN tandem was replaced in 1967 by a much larger tandem Van de Graaff accelerator with the HVEC designation of MP tandem. The latter, in turn, was decommissioned in early 1997.

Because of the pre-eminent role that tandem accelerators play in AMS research an account of their genesis and subsequent development is in order. The story began in the early 1930s with the invention of the Cockcroft–Walton generator [2] between 1930 and 1932 and the Van de Graaff electrostatic accelerator in 1931. These devices sparked a revolution in nuclear physics research. The article describing the Van de Graaff accelerator was published as a 16-line abstract of a paper delivered to a meeting of the American Physical Society [3]. The author was Robert J Van de Graaff, a soft spoken southerner born and raised in Tuscaloosa, Alabama, who, at the time, was a National Research Fellow at Princeton University. He later joined the faculty of the physics department at the Massachusetts Institute of Technology. There he and others founded HVEC to manufacture Van de Graaff accelerators of various sizes for a variety of applications. The terminal voltages (as described below) of these accelerators ranged up to about 5 million volts (MV). For many years their applications were confined to basic research in nuclear physics. Van was involved with the Manhattan Project to develop the atomic bomb during World War II and I recall listening to a radio drama, after the war, dealing with the development of the bomb. The actor playing the role of Van de Graaff had a pronounced guttural German accent in no way resembling Van's soft southern speaking voice.

Between 1930 and 1932 an alternative method for providing high d.c. voltages was proposed by Cockcroft and Walton and experiments involving the disintegration of elements by energetic protons were carried out [2]. Both methods are currently employed in tandem accelerators, especially for AMS. For simplicity I will concentrate on the Van de Graaff method of achieving high d.c. voltages.

After my discharge from the Royal Canadian Navy in 1945 I worked at Chalk River for a year as a research assistant. At that time the laboratory, called the Anglo-Canadian Atomic Energy Research Laboratory, was operated by the National Research Council of Canada. The director of the laboratory was J D Cockcroft of Cockcroft and Walton fame. He returned to the UK in 1946 to head the Atomic Energy Research Establishment

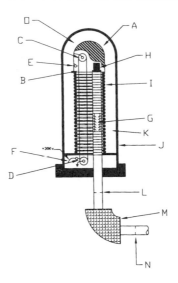

*Schematic diagram of a single-ended Van de Graaff accelerator: (A) high-voltage terminal—a hollow cylinder with a hemispherical cap; (B) rubberized fabric charging belt; (C) terminal belt roller and power generator; (D) belt roller at ground coupled to a drive motor; (E) electrons are removed from the terminal and deposited on the belt; (F) electrons are removed from the belt and deposited to ground; (G) acceleration tube; (H) positive ion source; (I) cage of ring electrodes; (J) pressure vessel; (K) insulating gas at a pressure of some 10 atmospheres; (L) entrance aperture to 90° magnet; (M) 90° energy analysing magnet; (N) exit aperture of 90° magnet; (O) corona points and generating voltmeter located in this vicinity between the pressure vessel and the terminal.*

in Harwell, England. That same year I enrolled as a graduate student in the Massachusetts Institute of Technology (MIT) where I earned my PhD in 1950. The roster of faculty in physics at MIT at that time, which included Van de Graaff, read like a *Who's Who* of American physicists. I returned to Chalk River, which had become a crown company called Atomic Energy of Canada Ltd, in 1952 and carried out research in nuclear physics using a Van de Graaff accelerator with a terminal voltage of 3 MV.

A schematic diagram of a vertical single-ended Van de Graaff accelerator is shown. It comprises a hollow metal cylinder with a hemispherical metal cap mounted on an insulated column. A rubberized fabric belt runs around a roller mounted at ground and around another roller inside the terminal. The belt roller at ground is rotated by a motor to which it is mechanically coupled. Electrons are removed from the insulated terminal and deposited on the belt. They are removed from the belt at the ground end. As this process proceeds a positive difference in voltage is established between the terminal and ground. This voltage difference can be controlled and varied as described below. A hollow, evacuated, insulated, vertical acceleration tube runs from inside the terminal to ground. An ion source capable of producing singly, positively charged atoms (hydrogen, deuterium or helium in the early accelerators) is mounted on the terminal end of the acceleration tube. Power needed in the terminal is supplied by a generator driven by the terminal belt roller. The column, ion source and tube structure are enclosed within a cage of equally spaced ring electrodes. The whole ensemble is mounted inside a pressure vessel containing dry nitrogen (N) or sulphur hexafluoride ($SF_6$) gas or a mixture of the two at pressures of ten or more atmospheres to insulate against sparking. Both the insulated column and the acceleration tube are constructed of layers of glass separated by thinner layers of metal. A high-resistance voltage divider runs from the terminal to ground with the individual metal layers of the column, the acceleration tube and the individual ring electrodes connected by resistors in this chain. This ensures that the terminal voltage is distributed uniformly from its maximum value at the terminal to ground along the tube, the column and the rings by the small current flowing through the resistor string. The terminal voltage can be readily varied. However, at a desired voltage, it can be held constant to a few parts in 10 000 or so as described below. The beam of charged particles accelerated by the terminal voltage passes down the acceleration tube and is highly collimated in the process.

The unit of energy used for charged particle beams in physics is the electron-volt (eV). It represents the kinetic energy imparted to a particle with a net charge of one electron accelerated through a potential of one volt. These so-called single-ended Van de Graaff accelerators have terminal voltages that range up to 5 MV and thus produce beams of singly charged particles with energies up to 5 MeV. Until about 20 years ago or so the highest terminal voltage that could be maintained stably in an electrostatic accelerator was not much over 5 MV. Beyond that value, voltage breakdowns occurred either through the high-pressure tank insulating gas or down the column or acceleration tube of the accelerator.

After reaching ground potential the path of the charged particle beam is bent from vertical to horizontal through 90° by a uniform magnetic field. The strength of the field in this magnet is measured in units of mass-energy product ($ME/q^2$) where $M$ is the mass number of the particle (one for protons, two for deuterons, four for alpha particles etc), $E$ is the particle energy in MeV where $V$ is the terminal voltage in units of a million volts and $q$ is the charge on the ion (one for hydrogen isotopes). In the case of singly charged ions $E = V$ and only those with the same value of the product of mass times energy $MV$ or $ME$ will be bent through $90°$ for a given magnetic field strength. This horizontal beam passes through a narrow slit between two metal plates, above and below the beam. If any part of the beam strikes a plate a current in a circuit attached to the plate is produced. If the terminal voltage is correct the beam passes centrally in the slit between the plates and any beam scraped off by the slit plates is equal at the top and bottom. If the terminal voltage rises the beam is bent less and produces an excess current in the lower plate. This is made to cause the terminal voltage to drop. Similarly if the terminal voltage drops the upper slit plate current increases and this signals the terminal voltage to increase. Changes in terminal voltage are accomplished by varying a corona current between the terminal and corona points inserted through the pressure vessel which are connected to a power supply. At the desired terminal voltage an equilibrium is established between the belt current, the column resistor current and the corona current. In this way the voltage on the terminal and hence the energy of the particles in the beam can be held constant to a few parts in 10 000. The net result is the production of a highly collimated, essentially monoenergetic beam of charged particles whose energy can be varied and which can then be used for nuclear physics research. The voltage between the terminal and ground is measured by a generating voltmeter. The latter device comprises a sectored rotor at ground potential which, as it revolves, alternately exposes and covers a signal plate to the field from the high-voltage terminal. The charging and discharging of the signal plate causes an alternating current to flow in an external circuit. This is rectified to produce a d.c. voltage proportional to the terminal voltage.

An enormous variety of research into the properties of atomic nuclei can be carried out with single-ended Van de Graaff accelerators. In general one wants to initiate nuclear reactions, which requires that the singly positively charged protons (for example) interact directly with the nuclei of the atoms in a target. Because the target nuclei are positively charged, as are the bombarding protons, the energy of the proton beam (and hence the terminal potential) must be sufficiently high to overcome the Coulomb repulsion or Coulomb barrier between particle and nucleus. For example for a 3 MeV proton (requiring a terminal potential of 3 MV) to penetrate to the surface of a target nucleus that nucleus must have a $Z$ (atomic number) of 18 or less, i.e. any element up to argon. If the terminal potential can be raised to 5 MV elements with $Z$ as high as 38 (strontium) or less can be studied.

Examples of the pioneering research possible with single-ended Van de Graaff accelerators are some experiments that suggested the application of an important nuclear model to light nuclei. A major programme at Chalk River among others, in the 1950s, was the study of the mass 25 isotopes of aluminium and magnesium for which the 3 MV Chalk River Van de Graaff machine was eminently suitable. In a publication in 1956 [4], it was pointed out that the collective model of Bohr and Mottelson [5] and its detailed application to the description of the properties of deformed nuclei by Nilsson [6], although originally conceived to explain the properties of nuclei in the rare earth and actinide regions of the periodic table, could also explain properties of nuclei in the region of mass 25. A comparison of the two mass 25 mirror nuclei was published in 1957 [7]. This work on mass 25 nuclei gained considerable encomiums for the Chalk River nuclear physics research groups and for nuclear physics research in Canada.

As time passed at Chalk River (and at other nuclear research laboratories throughout the world) it became obvious that one needed higher-energy beams. It is a fact of life in both nuclear and particle physics that ever higher energies are needed by researchers. In nuclear physics at that time (the mid 1950s) questions were being asked regarding the proton energy required to penetrate the Coulomb barrier of the heaviest element in the periodic table, namely uranium with an atomic number of 92. The answer is about 10 MeV. In an unpublished document dated 1 November 1954 entitled 'The place of a 10 MeV accelerator in a nuclear physics research programme' a colleague of mine, F B Paul, of the Nuclear Physics Branch at AECL at Chalk River, gave arguments for acquiring a Van de Graaff, as opposed to a cyclotron, capable of accelerating protons to an energy of 10 MeV. Paul had in mind a single-ended machine as described above with a terminal voltage of 10 MV. At that time, however, the technology was such that High Voltage Engineering Corporation (HVEC), which was the sole supplier of large electrostatic accelerators, was not willing to tackle such a high-voltage machine. The engineers and scientists at HVEC suggested that, instead, it be a tandem accelerator with a 5 MV terminal capable of producing 10 MeV protons as described below. It is interesting that, because of a long association between W B Lewis who was then vice-president for research at AECL and Denis Robinson who was president of HVEC, only a simple purchase order was required. Such an accelerator was designed, built and tested [1, 8], delivered to AECL and commenced operation in February 1959.

As pointed out by Van de Graaff [8] 'the principles and techniques used in tandem accelerators have been originated and developed over many years and in various laboratories'. As outlined below, tandem Van de Graaff accelerators are essentially two single-ended Van de Graaff accelerators joined at the high-voltage terminal. The usual arrangement is a linear one in which one accelerator ranges from ground to the terminal and the other from terminal to ground. An insulated accelerator tube runs from ground potential through the positively charged terminal and then through the second accelerator back to ground potential. Negative ions are injected into one end of the acceleration tube, accelerated to the terminal, changed to positive ions and accelerated away from the terminal back to ground. In Van de Graaff's paper [8] he highlighted the two developments that led to the design of tandem accelerators. The first was the discovery that negative hydrogen ions were stable and the second was that such ions could be given a double acceleration by changing their charge from negative to positive in the terminal.

In 1936 Bennett and Darby [9] published the results of their work that showed that an extra electron could be attached to both neutral hydrogen and deuterium ions and that the resulting singly charged negative ions were stable. In a paper published in 1953 [10] Bennett stated that, in 1935, he 'conceived the idea of using negative ions in a double high voltage tube for producing higher energy bombarding particles than accelerators with the same voltage in a single tube could produce'. He described this method in a 1937 patent which, in addition, included the important idea that, for elements of higher $Z$ than hydrogen, the singly charged negative ions could be converted to multiply charged positive ions when they interacted with the atoms in a thin foil or a differentially pumped gas cell in the terminal. They would thus gain an additional energy of $qV$ in units of electron-volts (eV) during the second acceleration where $q$ is the number of electrons stripped from the neutral atom and $V$ is the terminal voltage. The number of electrons stripped from the neutral atom $q$ is, obviously, less than or at most equal to $Z$. He noted [10] that 'nothing further was done with this idea until the hydrogen ion part of it was reinvented recently' citing L W Alvarez.

That reinvention occurred in 1951 [11], 14 years after the Bennett patent was issued when Alvarez, quite unaware of its existence, realized that stable singly charged negative ions of hydrogen could be accelerated to a high-voltage terminal, converted there to singly charged positive ions and then be further accelerated away from the same terminal. In his paper Alvarez gave a reference to a *Physical Review* article by Bennett (Bennett W 1936 *Phys. Rev.* **49** 91)—a reference that does not exist! Presumably it was meant to refer to the work of Bennett and Darby on the production of negative ions of hydrogen isotopes appearing further on in that same volume of *Physical Review*. In any case the Alvarez paper came to the attention of HVEC and made it possible, in 1954, for them to suggest that the way to accelerate hydrogen ions (both protons and deuterons) to energies of 10 MeV was to employ a tandem electrostatic accelerator with only a 5 MV terminal voltage. In a paper [12] delivered to the second AMS symposium in 1981 Alvarez gave a colourful description of how he invented accelerator mass spectrometry and the tandem accelerator among other things. In connection with the tandem he remarks in this paper that 'the man whose name is always associated with the tandem—Robert Van de Graaff— was not another independent inventor, but rather an excellent developmental engineer and Chief Scientist at the High Voltage Engineering Co. Neither Willard H. Bennett, who applied for a patent on the tandem in 1937, nor I, who first published it in 1951, are remembered as having anything to do with the matter'. It should be emphasized, however, that in Van de Graaff's paper [8] he referenced the negative ion work of Bennett and Darby [9], the Bennett paper of 1952 and the Bennett patent [10] and Alvarez' paper [11]. Alvarez went on to state [12] that, as soon as his 1951 paper [11] appeared, he received a phone call and later a visit from his old friend from wartime days, Denis Robinson, then president of HVEC. Robinson said he wanted to build and sell tandem accelerators based on the principle invented by Alvarez. As stated above such an accelerator was designed, built, tested and delivered to AECL. It commenced operation for nuclear physics research in February 1959.

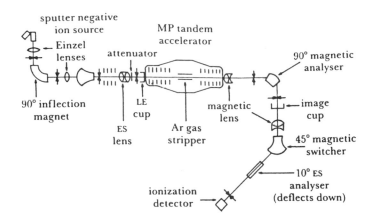

*Schematic diagram of an MP tandem Van de Graaff AMS system.*

A schematic of an MP tandem Van de Graaff accelerator is shown. As mentioned previously they are essentially two single-ended Van de Graaff accelerators mounted collinearly head-to-head with a common positive high-voltage terminal. The ion source, which is conveniently located outside the pressure vessel, produces both singly positively and negatively charged ions as well as neutral atoms of whatever element has been loaded into the source. Further details of the ion source are given later.

An analysing magnet selects negative ions from the source for injection into the first half of the accelerator. They are attracted to the positively charged terminal arriving there with an energy of $V$ million electron-volts (MeV). Again $V$ is the voltage on the terminal in units of a million volts. In the terminal they pass through a so-called stripper comprising either a thin foil (usually carbon) or a differentially pumped tube into which gas (usually argon) is injected. The interaction of the energetic negative ions (they reach energies of several MeV in the first stage of their acceleration) with the atoms in the stripper results in the removal not only of the extra electron that made the atom negative but of several more electrons causing it to have a multiple positive charge. If $q$ such electrons are removed from the neutral atom (where $q$ is less than or equal to $Z$, the atomic number of the accelerated atom) the ion receives an additional energy $qV$ through the second acceleration stage.

The ions arrive at ground potential with a total energy of $(1+q)V$. They are then deflected through 90° in an analysing magnet. The beam of ions, usually amounting to several microamperes of current, passes through a narrow slit between two metal plates. The difference in current striking these two plates provides a signal to stabilize the voltage on the terminal just as in the case of the single-ended Van de Graaff accelerator described previously. Again the strength of this analysing magnet is measured in units of its mass energy product ($ME/q^2$). Now the energy is $(1 + q)V$ so the magnetic field must have a mass energy product of $MV(1 + q)/q^2$. So, for example, a beam of fluorine ions, of mass 19, in which four of the nine electrons of each ion were removed in the stripper, with 5 MV on terminal requires a magnet with a mass energy product of approximately 30 ($19 \times 5 \times (1 + 4)/16$) to bend the beam through 90°. In the stripper the particles in the beam divide into a number of charge states whose distribution is described by a statistical model as a bell-shaped curve. For example, if the stripper is a differentially pumped tube of argon gas, fluorine atoms of energy 5 MeV, after passing through the stripper have 46% of the ions in charge state 4, 24% in charge states 3 and 5 and only about 3% in charge states 2 and 6. For a foil stripper the charge states are one unit higher.

Following the analysing magnet the ion beam passes into a switching magnet which can be set to send it into any one of several beam lines. Each beam line terminates in a target in which the beam particles initiate nuclear reactions with atoms in the target. The reaction products are then measured in a variety of specialized apparatus depending on the nuclear properties that the experiment is designed to investigate.

As mentioned previously, the first tandem Van de Graaff accelerators were designated by HVEC as EN tandems. They had a nominal terminal voltage of 5 MV. Over two dozen such machines were manufactured and employed for nuclear physics research throughout the world, the first by nuclear physicists at Atomic Energy of Canada Ltd starting in 1959. Once again, however, the demand for higher energies increased the pressure on HVEC to produce a higher-voltage tandem accelerator. Their response was the FN tandem and its upgraded version, the Super FN, with terminal voltages of 7.5 and 9.0 MV respectively. Some 17 of these were produced. The next accelerator in the series was the MP tandem Van de Graaff with terminal voltage of 10 MV. Through the years a variety of engineering upgrades allowed the MP voltage to be increased to 13 MV and higher. The first MP tandem was installed at Yale University in 1965, the second and third at the University of Minnesota and at Atomic Energy of Canada Ltd that same year and the fourth at the University of Rochester a year later. It was this latter accelerator that was employed for the first direct detection of carbon-14 in natural organic samples in 1977 (reference [9], Chapter 2). Six more MP tandems were manufactured and installed in nuclear physics laboratories, two each in Germany, USA and France.

A more detailed account of the development of electrostatic accelerators has been published by Bromley [13]. This issue of *Nuclear Instruments and Methods* (NIM), edited by Bromley and published before the invention of AMS, was devoted to 'large electrostatic accelerators' and contained 21 articles on various aspects of such accelerators. In Bromley's article [13] reference is made to two 5 MV tandems built in Japan by Toshiba and Mitsubishi and installed in the late 1960s at the Universities of Tokyo and Kyoto respectively as well as a comparable tandem designed and built in the USSR. Also included are references to two double tandem installations built by HVEC at the Universities of Pittsburgh and Washington and the double MP tandem installation at Brookhaven National Laboratory. The contributions to the design and manufacture of high-voltage electrostatic accelerators made by the National Electrostatic Corporation (NEC) in the USA are also described. These include an 8 MV tandem installed in Sao Paulo, Brazil in 1970 and two 14 MV tandems installed at the Australian National University in Canberra in 1972 and some time later at the Weizmann Institute in Israel. Other notable high-voltage electrostatic accelerators mentioned [13] include the HVEC 20 MV XTU in Legnano, Italy and the 30 MV Nuclear Structure Facility tandem designed and built in-house at the Daresbury Research Laboratory in the UK. A folded or up–down 25 MV tandem built by NEC and installed at the Oak Ridge National Laboratory has been in operation for several years, as has a home-built smaller folded tandem at Oxford University in the UK and at the University of Auckland in New Zealand. The MP tandem at the C.R.N. (Centre de Recherche Nucléaire) in Strasbourg, France was converted to a much larger tandem called a Vivatron. Unfortunately it has not yet achieved the 30 MV or so terminal voltage its designers had hoped for.

The success of tandem accelerators for both nuclear physics research and for AMS is critically dependent on the design of ion sources capable of producing copious beams of negative ions of elements throughout the periodic table. One of the major contributors to the science of negative ion sources, Roy Middleton, published an article in the above issue of NIM on such sources [14]. He later published *A Negative-Ion Cookbook* [15] which rapidly became required reading for users of tandem electrostatic accelerators.

In the NIM issue cited above, Middleton [14] noted that there were four methods for producing negative ions in common usage at the time. These were charge exchange, high-voltage dissociation, direct extraction and sputtering. In the case of the charge exchange source, positive ions of the species desired as negative ions are accelerated to energies of a few tens of keV. They pass through a donor medium of gas or vapour and some 1% or so emerge as negative ions. Donor media of hydrogen gas, cesium or lithium vapours have been employed. The positive ions can be obtained from a variety of positive ion sources. In high-voltage dissociation sources a 40 keV primary beam of krypton or xenon passes through a gas or vapour medium. A small fraction of the molecular components of the medium form negative ions which can be extracted and injected into the low-energy end of the tandem. Direct extraction sources are standard positive ion sources in which the extraction polarity is reversed. In 1959 Moak *et al* showed that negative ion currents could be extracted directly from a widely used positive ion source called a duoplasmatron [16]. The basic principles of the sputter source were described by Middleton and Adams in 1974 [17]. It utilized a 30–35 keV beam of positive cesium ions to bombard a solid target of the neutral atoms to be transformed to negative ions. The cesium beam performs two functions. It forms a thin layer of neutral cesium on the target surface which reduces the work function and improves the negative ion yield and it sputters the neutral atoms out of the target through this cesium layer. For the next decade it was sources of this type or variants thereof that were employed on most tandem accelerators. Middleton described an improved version of such a sputter ion source in 1983 [18]. In the ion sources described in this paper the two functions of the cesium, to provide a coating of neutral cesium atoms on the target surface and to sputter atoms from the target, are separated and very much higher negative ion currents are produced.

# Chapter 5

# Instrumentation for Accelerator Mass Spectrometry

We turn now to the use of tandem accelerators for accelerator mass spectrometry; a number of review articles have been published on the subject beginning in the late 1970s [1–7]. These reviews emphasize the advantages of AMS over direct counting methods in the detection of atoms of trace elements, both stable and radioactive, in matrices of much more abundant elements. These advantages of AMS are a consequence of the use of negative ion sources, the acceleration of ions to MeV energies, the elimination of molecules by multiple electron stripping in the terminal and the use of well known nuclear physics techniques to determine the atomic mass and atomic number of the elements to be measured.

The major use of AMS until now has been in the ultrasensitive detection of certain long-lived radionuclides. The five most common of these are beryllium-10 ($^{10}$Be), carbon-14 ($^{14}$C), aluminium-26 ($^{26}$Al), chlorine-36 ($^{36}$Cl) and iodine-129 ($^{129}$I) with half-lives of $1.6 \times 10^6$, 5730, $7.05 \times 10^5$, $3.01 \times 10^5$ and $1.57 \times 10^7$ years respectively. The production modes and the applications of these five radioisotopes are discussed below.

## Beryllium-10

Beryllium-10 is mainly produced in the atmosphere by nuclear reactions initiated by high-energy cosmic ray particles striking nitrogen-14 and oxygen-16 atoms and causing several particles to be emitted. The process is referred to as spallation. Beryllium-10 attaches itself to particulate matter in the atmosphere and is precipitated to earth in rain or snow.

It has been measured by AMS to give information about the terrestrial age of meteorites, to calibrate the cosmic ray flux using ice and sediment cores and to provide information on erosion and sedimentation rates. Its measurement by AMS has many geological applications including rock surface exposure ages and information concerning weathering of geological features and the advances and retreats of glaciers.

## Carbon-14

Carbon-14 is produced in the atmosphere mainly by low-energy cosmic ray neutrons interacting with nitrogen-14. The nuclear reaction involved is $^{14}N(n, p)^{14}C$ in which nitrogen captures a neutron, emits a proton and carbon-14 is formed. Some carbon-14 is produced by the spallation of oxygen-16 in the atmosphere. In both cases the carbon-14 quickly reacts with oxygen in the air to form $^{14}CO_2$. Living organic material ingests $CO_2$ and $CO_2$ is also transported to the oceans by ocean–air interactions.

The main and most fascinating application of carbon-14 is in archaeology and anthropology where the measurement of its ratio to that of stable carbon establishes the time at which an organic system died. It can also be used to measure the terrestrial ages of meteorites, the age of ground water and to calibrate cosmic ray flux using tree rings, ice and sediment cores and coral. It can also be used to study ocean and atmospheric circulation and provide information on carbon compounds in the air. Like beryllium-10, there are many geological applications of AMS studies of carbon-14. Carbon-14 is the most important radioisotope in biological and medical studies.

## Aluminium-26

Aluminium-26 is mainly produced in the atmosphere by the spallation of argon-40 by high-energy cosmic rays. Like beryllium-10, it becomes associated with particulate matter in the atmosphere and precipitates to earth as rain and snow.

It is used to measure the terrestrial ages of meteorites. Like beryllium-10, it has many geological applications. It also has some medical applications.

## Chlorine-36

Like aluminium-26 above, chlorine-36 is mainly produced in the atmosphere by spallation reactions on argon-40. There are negligible contributions from two other nuclear reactions, $^{36}Ar(n, p)^{36}Cl$ and $^{35}Cl(n, \gamma)^{36}Cl$. It precipitates to earth in rain or snow.

Next to carbon-14, chlorine-36 has the most applications. It is used to measure the ages of ground water and the terrestrial ages of meteorites, to trace the movement of ground water, to trace the leakage of nuclear waste and to calibrate the cosmic ray flux using ice cores. It has also been used to measure the neutron flux from the nuclear bombs used on Hiroshima and Nagasaki. It has many geological applications, even more than the three previous radioisotopes. Of these an important one is the dating of the exposure ages of rocks.

## Iodine-129

Iodine-129 is produced in the atmosphere by the spallation of xenon. It is also produced in the earth by the spontaneous fission of uranium. Its anthropogenic production comes mainly from nuclear fission in reactors and nuclear weapons.

It is used to trace the leakage of nuclear waste from storage facilities and the production of fission products from nuclear weapons explosions and nuclear reactor accidents. It can be used to date sediments, trace slow marine water movements and to date and trace hydrocarbons, particularly oil.

The applications of AMS in the measurement of the most important radionuclides listed above are the most common ones and fail to reflect the enormous variety of measurements encompassed in each category. In particular the archaeological, environmental science and geoscience applications are staggering in their variety. Not listed in the above are applications in atomic, nuclear and particle physics and materials science in which AMS has played and will continue to play a key role.

There are several other radioisotopes and stable isotopes whose concentrations at levels of parts per trillion in matrices of other elements can be measured by AMS and whose measurements are of considerable value. The former include silicon-32 and calcium-41 with half-lives of about 150 and $10^5$ years respectively and the latter boron, phosphorus, antimony, iridium and platinum. The applications of such measurements will be discussed later in the book.

It is fair to say that AMS places more stringent demands on the performance of tandem accelerator systems than does their use for nuclear physics research. The basic reason is that, for AMS, tandems are employed to detect isotopes in samples that occur at concentrations of a few parts in $10^{15}$. For example, in the field of carbon dating, an organic sample that 'died' some 60 000 years ago has a ratio of carbon-14 atoms to stable carbon-12 atoms of about $10^{-15}$, while for a modern living sample the ratio is around $10^{-12}$. In addition to measuring this ratio, a well designed tandem accelerator system for carbon dating by AMS measures the ratio of the two stable carbon ions, carbon-13 and carbon-12. This ratio should be close to 1%. For ratios of carbon-14 to carbon-12 and carbon-13 to carbon-12 in present-day organic samples one often requires accuracies or reproducibilities in the measurement of the ratio of 0.5% or less. Such extreme accuracy is virtually never required of measurements, even of ratios, in nuclear physics research.

Another way of expressing the problems involved in tandem AMS measurements of, for example, carbon is to consider the beam currents involved. Beam currents of the stable isotope carbon-12 from tandem accelerators can range from 1 to 50 microamperes. This translates into $6 \times 10^{12}$ to $300 \times 10^{12}$ carbon-12 atoms per second. Beam currents this high provide plenty of signal, when passing through the narrow slit following the $90°$ analysing magnet after acceleration through the tandem, to stabilize the tandem terminal voltage to the requisite accuracy. The comparable carbon-13 beams are 0.01 to 0.5 microamperes of carbon-13 or $6 \times 10^{10}$ to $300 \times 10^{10}$ carbon-13 atoms per second, and even these beams are sufficient to provide slit stabilization of the terminal voltage. However, if the carbon beam is from a modern organic carbonaceous sample with a carbon-14 to stable carbon ratio of $10^{-12}$, the intensity of carbon-14 atoms is 6 to 300 particles per second and for a 60 000-year-old sample the intensities are 1000 times lower. In one version of tandem systems for AMS the whole system from ion source to final detector is sequentially tuned to accept carbon-12, carbon-13 and carbon-14. The intensities of the latter are much too small to provide slit stabilization and another method of keeping the terminal voltage constant must be employed.

In all applications of tandem electrostatic accelerators the terminal voltage is measured by a device called a generating voltmeter (GVM), described in the previous chapter. It produces a d.c. voltage proportional to the voltage on the terminal. In the AMS mode, when the beam current is too low to provide slit stabilization, this GVM voltage signal can be compared with an external voltage that can be varied, and differences in the two voltages can be arranged to provide a feedback system to stabilize the terminal voltage. GVM stabilization has been developed to permit terminal voltage control with beams of tens to hundreds of particles per second and the voltage stability achieved in this way is comparable with that provided by slit stabilization for beam currents of a fraction to tens of microamperes.

The precision with which AMS measurements of, for example, carbon-14 atoms from a sample can be made depends on a number of factors. Among these an important factor is counting statistics. It is well known that if $N$ events are counted and if the frequency distribution of these events is governed by Poisson statistics (as are most statistical processes), the standard deviation is $N^{1/2}$ [8]. This means that, if the required counting statistical accuracy is 0.5% or ±41 years, 40 000 atoms of carbon-14 must be collected from the sample. For carbon from a modern organic sample and an output carbon-12 beam from the accelerator of 50 microamperes it only takes 2.2 minutes to accumulate that many carbon-14 counts. It should be remarked that, although the half-life of carbon-14 is 5730 years, the mean life is 8270 years and it is the latter to which the 0.5% must be applied ($0.5 \times 10^{-2} \times 8270 = 41$).

There are essentially five elements to an AMS system based on tandem accelerators (see, for example, [6]), the ion source, the injector, the tandem accelerator, the positive ion analysis system and the detection system. The details of each of these systems varies somewhat from one AMS laboratory to another.

Most AMS systems currently employ a cesium sputter source and some details of such sources have been discussed in Chapter 4. The samples or targets used in these sources are solids; in the case of carbon-14, for example, a variety of techniques have been devised for converting small organic samples into solids suitable for use in a sputter source (see, for example, reference [6], Chapter 3). Most AMS carbon dating laboratories prefer to convert

ASSUME CONTEMPORARY ORGANIC CARBON WITH C-14 / C = 1.2 X 10⁻¹²

| ACCURACY | ERROR | COUNTING TIME DECAY | AMS* |
|----------|-------|---------------------|------|
| 10% | ±830 y | 7 m | 0.0 s |
| 1.0% | ±83 y | 12 h | 27 s |
| 0.1% | ±8 y | 1200 h | 44 m |

* ASSUMING A CARBON ION SOURCE CURRENT OF 50 X 10⁻⁶ AMPERES

T1/2 = 5730 YEARS FOR C-14  HALF LIFE

T$_m$ = 8270 YEARS FOR C-14  MEAN LIFE

DECAY COUNTING REQUIRES ABOUT 1 GRAM OF CARBON

AMS REQUIRES ABOUT 1 MILLIGRAM OF CARBON

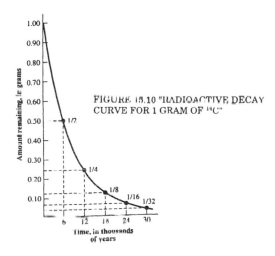

FIGURE 15.10 "RADIOACTIVE DECAY CURVE FOR 1 GRAM OF ¹⁴C"

*Comparison between decay counting and AMS for radiocarbon dating and information on the accuracy of radiocarbon dating measurements.*

the organic sample to $CO_2$ gas and then crack the gas to graphite, since graphite produces the highest-intensity beams of negative carbon ions. There would be advantages to inserting the $CO_2$ gas directly into an appropriate ion source. The problem is that, so far, contamination in the ion source from previous samples (the so-called memory effect) has, for ion sources employing gas, been found to be difficult to overcome. For radioisotope measurements sample sizes of 1 to 10 milligrams are employed and, generally, several samples can be mounted for insertion into the source without disturbing the vacuum system involved. As mentioned above the five long-lived radionuclides measured by AMS are

**Table 5.1.** *Typical parameters for the five radionuclides most commonly measured by AMS in natural samples.*

|  | $^{10}$Be | $^{14}$C | $^{26}$Al | $^{36}$Cl | $^{129}$I |
|---|---|---|---|---|---|
| Half-life (years) | $1.6 \times 10^6$ | 5730 | $7.05 \times 10^5$ | $3.01 \times 10^5$ | $1.57 \times 10^7$ |
| Stable isotopes | $^9$Be | $^{12}$C, $^{13}$C | $^{27}$Al | $^{35}$Cl, $^{37}$Cl | $^{127}$I |
| Stable isobars | $^{10}$B | $^{14}$N$^{\#}$ | $^{26}$Mg$^{\#}$ | $^{36}$Ar$^{\#}$, $^{36}$S | $^{129}$Xe$^{\#}$ |
| Chemical form | BeO | C | Al$_2$O$_3$ | AgCl | AgI |
| Terminal MV | 7, 8 | 2 | 7.5 | 8, 10 | 5 |
| Charge state | 3 | 3 | 7 | 7 | 5 |
| Energy (MeV) | 28, 32 | 8 | 60 | 64, 80 | 30 |
| Sample size (mg) | 0.2 | 0.25 | 3 | 2 | 2 |
| Bkgnd. ($\times 10^{-15}$)$^{\#\#}$ | 3 | 0.5 | 1 | 0.5 | 10 |
| Ion source (IS) current ($\mu$A) | 12 | 7 | 1.5 | 8 | 3 |
| Stripper yield | 0.54 | 0.42 | 0.35 | 0.32 | 0.10 |
| Accel. efficiency | 0.28 | 0.71 | 0.43 | 0.18 | 0.23 |
| Run time (min)* | 5 | 2 | 120 | 30 | 20 |
| Count time (years)** | 570 | 1 | 150 | 60 | 1200 |
| Decay/AMS*** | $6 \times 10^7$ | $2.5 \times 10^5$ | $7 \times 10^5$ | $1 \times 10^6$ | $3 \times 10^7$ |

# Isobar does not have a stable negative ion.
## The background is the ratio of the radioisotope to the stable isotope for a sample with a negligible radioisotope content prepared in the same way as the unknown sample.
* The time required to count 100 radioactive atoms in the final AMS detector for a sample that has an isotope ratio an order of magnitude higher than the background value.
** The time required to count 100 radioactive decay events for the same sample size as given in row 8 in the above table placed in a detector that is 100% efficient.
*** The ratio of time required to count 100 radioactive decay events to the time required to count 100 radioactive atoms by AMS.

beryllium-10, carbon-14, aluminium-26, chlorine-36 and iodine-129. The chemical forms generally used for each are BeO, carbon (graphite), Al$_2$O$_3$, AgCl and AgI respectively (see table 5.1). The negative ions emerge from the source with energies of 25 keV or so.

Prior to insertion of these ions into the first half of the tandem accelerator it is advantageous to provide as high a mass resolution analysis as economically feasible using a magnetic analyser. A mass resolution of $M/\Delta M = 300$ or higher is desirable. With such resolution the isotopes of the heaviest elements can be resolved. The addition of an electrostatic analyser preceding the magnetic analyser is used in some installations. It reduces the tails of the energy distribution of the ions produced from the cesium sputter ion sources. The negative ions are then further preaccelerated to energies up to a few hundred keV depending on the size of the tandem and injected into the first half of the tandem focused on the stripper in the centre of the high-voltage terminal. The stripper is, typically, a differentially pumped tube into which argon gas is fed.

The terminal of the tandem is held at a constant positive voltage ranging from 2 to 10 MV. In the above list of five radioisotopes only chlorine-36 requires the highest terminal voltage, as will be explained later. In the terminal stripper, in the interactions between the energetic negative ions being accelerated and the stripper gas, not only is the extra electron that made the accelerated ions negative removed but, as described above, additional electrons are stripped off creating multicharged positive ions. The principal sources of interference in AMS will be discussed later. As will be seen, molecular interference is eliminated, in general, when three or more electrons are removed from a neutral molecule. The multiply charged positive ions then undergo a second acceleration from the terminal to ground. They arrive there with an energy of $E = (1 + q)V_T + V_I$ MeV where $q$ is the number of electrons removed from the neutral atom in the terminal gas stripper, $V_T$ is the terminal voltage in MV and $V_I$ is the injection voltage before entering the tandem.

Positive ion analysis systems vary from one AMS laboratory to another. All of them, however, include both magnetic and electrostatic fields. The former select $ME/q^2$ and the latter $E/q$ where $M$ is the ion mass, $E$ its kinetic energy and $qe$ its charge. Another analysis system that is sometimes used is a Wein filter. This comprises crossed electric and magnetic fields adjusted so that ions of a specific velocity pass through undeflected. Such a device selects $E/M$. The measurement of the time-of-flight of an ion also selects ions of a specific $E/M$. Magnetic or electric deflection systems alone are not sufficient because it is necessary to uniquely define both $M/q$ and $E/q$ such that only particles having this parameter combination can reach the detector [2]. It has been noted [2] that even this does not provide a unique identification. Multiples and submultiples of the variables $M$, $E$ and $q$ represent classes of particles that cannot be distinguished; molecules having a common multiplier of $M$, $E$, $q$, ignoring mass defects, represent another. Some AMS systems employ a second magnet at the high-energy end to reduce interference from scattering and charge changing due to interactions with residual gas in the vacuum lines. The University of Toronto system is shown.

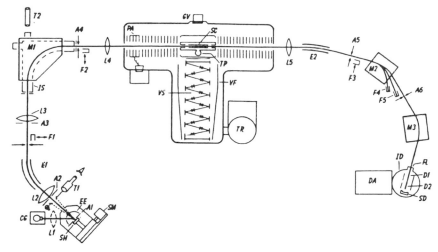

Schematic layout of the IsoTrace Tandetron AMS facility. No steerers are shown. Ion source: CG-cesium gun; L1-split einzel lens; SH-sample holder; SM-sample motion; EE-electric extraction; A2–phase space defining aperture; L2–einzel lens. Low-Energy-Mass Analysis: E1-electric analyzer; A3-phase space defining aperture; L3-einzel lens; M1-injection magnet; IS-electric isotope selector ("bouncer"). Accelerator: L4-matching einzel lens; PA-acceptance matching lens; SG-argon stripper gas cell; GV-rotating turbo pump; VS, TR and VF-accelerator power supply. High-Energy-Mass Analysis: L5-electric quadrupole lens; E2-energy and charge state defining electric analyzer; M2 and M3-magnetic analyzers; F4 and F5-$^{12}$C and $^{13}$C-faraday cups; ID-ionization detector.

*Schematic diagram of the IsoTrace Tandetron AMS facility at the University of Toronto. Courtesy of the IsoTrace Laboratory, University of Toronto.*

The heaviest radioisotope that is commonly measured in AMS systems is iodine-129. In recent years, however, increasing attention has been paid to the actinides (the 15 elements beginning with actinium, atomic number 89). The masses of interest for these latter elements range from about 230 to 245. The key characteristics of the components in the positive ion analysis systems at the high-energy end of an AMS facility for such high masses and even for iodine-129 are the mass energy product $ME/q^2$ of the analysing magnet and the strength $E/q$ of the electric analysers. The highest mass energy product 90° magnet currently available on AMS systems has a value of about 155. In the case of $^{129}$I the charge state selected in the terminal is $q = 5$ and the terminal voltage must be no higher than 5 MV. This gives $ME/q^2 = M(1 + q)V/q^2 = 155$. For the actinides the terminal voltages required are generally limited to about 2 MV. For example for uranium-236, whose ultrasensitive detection by AMS is of considerable interest, for such a terminal voltage a charge state $q = 4$ must be selected from the terminal stripper. There is an efficiency of about 10% for producing such a charge state using argon gas in the terminal stripper. Higher mass energy product magnets would allow the selection of more efficiently produced higher charge states. At least two AMS laboratories employing 2 to 3 MV tandems (tandetrons) will have 90° analysing magnets with much higher mass energy product magnets in the near future.

A variety of detectors are employed following the magnetic and electric deflection analysers at the high-energy end of the AMS systems. The most common for the lightest radioisotopes such as carbon-14 is a $\Delta E - E$ Si surface barrier solid state detector. As detailed below, the combination of these two energy signals defines both $M$ and $Z$.

For intermediate-mass radioisotopes such as chlorine-36 a multianode gas ionization detector is employed. The ions enter the detector through a thin window and pass through the counter gas producing a track of ionization. Such a detector measures the decrease in energy $dE$ usually in three sections $dx$ of path along the trajectory and the residual energy loss in the remaining part of the ions' path. The summed decrease in energy in the three sections when added to the residual energy loss defines the energy $E$ of the ion. The energy loss, $dE/dx$, is proportional to $MZ^2/E$ where $M$ and $Z$ are the ion's mass and atomic number respectively and $E$ is its energy at any point along its path in the detector's gas. Thus if both $E$ and $dE/dx$ are measured it is possible to discriminate between isobars (ions of the same mass $M$ but different $Z$).

This is of crucial importance in the case of chlorine-36 ($Z = 17$) where the stable isotope of sulphur, sulphur 36 ($Z = 16$), is the interfering isobar. In order to traverse gas ionization chambers of suitable operating design, however, the ions must have a minimum energy. In the case of mass 36 this turns out to be about 60 MeV (although even higher energies are preferable), requiring tandem accelerator terminal voltages of at least 8 MV. Thus, of all the atoms detected by AMS, only chlorine-36 requires tandems with terminal voltages much higher than about 2 MV. In addition, careful chemical procedures must be carried out to reduce the sulphur content of the samples. Otherwise the counting rate of sulphur-36 in the gas ionization chambers would exceed the maximum rate such devices could accommodate.

Time-of-flight measurements can be employed to identify the mass–energy ratio of ions and thus to distinguish them from the tails of unwanted ions that manage to pass the magnetic and electric deflectors on the high-energy line. Such detectors are typically 1 to 3 metres long. When an ion enters the device it initiates a start signal by causing the emission of secondary electrons as it passes through a thin carbon foil at the entrance. These electrons are deflected into a microchannel plate detector generating a fast electronic pulse. At the end of the flight path a stop signal is generated in a similar fashion or by an Si surface barrier detector. The time difference between the start and stop signal is proportional to $(M/E)^{1/2}$.

Such a device is of particular value in the case of iodine-129. In this case there is no interfering stable isobar (the negative ion of xenon-129 is unstable). However there is an unwanted ion, namely the much more intense stable isotope of iodine, $^{127}$I. Although it is two units of mass less than $^{129}$I a combination of charge exchange and scattering causes tails of iodine-127 to enter the final detector following the time-of-flight system along with the wanted iodine-129. Because they have the same rigidity as the iodine-129 ions their energy is 1.5% less and their flight time through the device is also 1.5% less. This enables them to be distinguished from iodine-129 (see reference [19], Chapter 3).

AMS measurements consist of establishing the ratio of the counting rate of the radioisotope in question to the current of its stable isotope, i.e. $^{10}$Be/$^{9}$Be, $^{14}$C/$^{12}$C and/or $^{14}$C/$^{13}$C, $^{26}$Al/$^{27}$Al, $^{36}$Cl/$^{35}$Cl and/or $^{36}$Cl/$^{37}$Cl and $^{129}$I/$^{127}$I. Standards of known isotope ratio are run frequently for normalization and blanks with no detectable radioisotope are employed as a measure of background. Corrections for other effects can be made by comparison with the standards. Table 5.1 (after that in [6]) gives typical parameters for the above five radioisotopes.

The numbers given in the last row of table 5.1 are impressively large. In the case of carbon-14, they mean that a measurement can be made of an artefact 23 500 years old with a statistical accuracy of 10% ($\pm$830 years) in about 2 minutes using a 0.25 milligram sample, whereas for the same sample size it would take almost a year by decay counting to obtain the same accuracy. A modern sample could be measured to 1% ($\pm$83 years) in 11 minutes by AMS using a 0.25 milligram sample while, for the same sized sample, it would take 5 years by decay counting.

Some of the limitations of AMS have been described [6]. Unlike other types of mass spectrometers AMS does not have a high mass resolution. Generally the magnets following the negative ion source and prior to injection into the tandem accelerator have a mass resolution, $M/\Delta M$, of 300 or less. This is sufficient to separate isotopes, for example $^{236}U$ from $^{235}U$, but far too low to discriminate against atomic isobars (atoms of the same mass), for example $^{14}N$ in the case of $^{14}C$, $^{36}S$ in the case of $^{36}Cl$, etc or molecular isobars (molecules of the same mass), for example $^{12}CH_2$ or $^{13}CH$ in the case of $^{14}C$, $H^{35}Cl$ in the case of $^{36}Cl$, etc. In many cases negative ions of the atomic isobar are unstable, but if not one can take advantage of the fact that, although they have the same atomic masses they have different atomic numbers (differing by one atomic mass or more) and they can be minimized by careful chemistry and then further reduced or eliminated by $dE/dx$ measurements in an absorber. The latter requires tandems with terminal voltages above about 8 MV to produce high enough energies to make the $dE/dx$ differences for atoms of adjacent $Z$ sufficiently large. An AMS system was designed to have mass resolution of the ion source magnet as high as $M/\Delta M = 1000$ [9] but the system is no longer operating. Even this mass resolution is insufficient for the elimination of atomic or molecular isobar interference.

On the other hand molecules, in general, are not a problem. The main advantage of AMS is that molecules are destroyed in the terminal when the extra electron is stripped from the negative molecule to make it neutral and then three or more electrons are stripped from the neutral molecule. The resulting charge 3+ state or higher of a molecule is unstable and it blows apart by a so-called Coulomb explosion.

There are other sources of interference, generally referred to as background [6]. The chief backgrounds are contamination of the sample up to and including insertion into the ion source by the environment, cross contamination (memory) from previous samples in the ion source, scattering from surfaces, charge changing from collisions with residual gas in the vacuum and unresolved isobars and isotopes. All these contribute to the background listed in table 5.1. For most applications of AMS the ratio of the radioisotope to the stable isotope in the unknown sample is appreciably higher than the background listed in table 5.1.

Various factors can affect the precision of an AMS measurement [6], the most important of which is counting statistics referred to previously. The precision of most AMS measurements lies between 3 and 10%. Carbon-14 is an exception. Here a well operated AMS facility can achieve a precision of 1% (±83 years) and, if pressed and the sample warrants, 0.5% or 5 per mil (per thousand) (±41 years)

As mentioned previously there were some 40 full time or part time AMS laboratories throughout the world by early 1997 [10]. There are 41 listed in [10] but one of these, that at Chalk River, was decommissioned a few months after that article was published. Of the remaining 40, all but two employ tandem electrostatic accelerators. One of these two is located in Shanghai, China and is based on a 50 keV negative ion minicyclotron. It was designed exclusively for radiocarbon dating. Its success as a competitor to small tandem electrostatic accelerators is still problematical. The second is the major nuclear physics research accelerator facility at Michigan State University, based on two large superconducting cyclotrons. Its use in AMS measurements is minimal. Of the 38 AMS facilities employing tandem electrostatic accelerators, sixteen have terminal voltages between 2.5 and 3 MV, ten between 5 and 6 MV, six between 8 and 9.5 MV and six between 12 and 20 MV. Except for AMS measurements of chlorine-36 and calcium-41 which require tandems with terminal voltages of 8 MV or higher (to resolve the corresponding stable isobars sulphur-36 and potassium-41) all other light and heavy radioisotopes and stable isotopes can be readily measured with tandems having terminal voltages as low as 2.5 MV.

# Chapter 6

# The Legacy of the Atomic Bombing of Hiroshima and Nagasaki

I was born in Niagara Falls, Ontario, Canada and received my primary and secondary school education there. In the fall of 1940 I entered Queen's University in Kingston, Ontario as an undergraduate in engineering physics. One year before, on 1 September 1939, Germany invaded Poland and two days later Britain declared war on Germany. Canada's declaration of war followed without delay. There was no draft in Canada and those of us at university or about to enter university were encouraged to continue our education.

By mid October 1939, three weeks after the fighting in Poland was over, Britain had four divisions comprising 158 000 soldiers in France—most of its fighting force at the time. France on its part had almost 40 divisions and yet, despite its promises to Poland, refrained from taking any action against a greatly inferior German force on the Franco-German border. There followed what was called in the West 'the phony war' during which Germany conquered Denmark, Norway, Holland and Belgium and in 1940 trapped the British and some of the French forces in an area surrounding Dunkirk. The only hope was an evacuation across the channel. In May 1940 the British Expeditionary Force along with substantial numbers of French, a total of some 340 000 in all, were evacuated from Dunkirk by a seeming miracle. A month later France collapsed.

I began my university career at the height of the Battle of Britain when the German *Luftwaffe* attempted to bomb Britain to its knees

in preparation for an invasion. Only the indomitable spirit of
the Royal Air Force thwarted Hitler's plans. This one bright light
flickered briefly to be followed by months of darkness and despair.
In June 1941 Germany invaded Russia. She was now the master
of Europe and seemed to me to be well on the way to becoming
master of the world.

All this time Canada's mighty neighbour to the south was in
a state of paralysed neutrality caused by the powerful advocates
of isolationism. This isolationist attitude was not shared by the
President of the United States, Franklin Delano Roosevelt, but
politically his hands were tied. In the fall of 1941 I began my
second year at Queen's. As I sat studying in my shabby rooming
house in Kingston on the morning of Sunday 7 December 1941
I knew what all of us in Canada knew, that the only hope for
the eventual defeat of Hitler depended on the entry of the United
States into the war. There seemed little prospect of that.

Some time after noon as I was listening to the radio the
programme was interrupted by a CBC announcer. Pearl Harbor
had been bombed by a large contingent of the Japanese Air Force.
Major elements, perhaps virtually all, of the US Pacific Fleet had
been sunk! It was a disaster of unparalleled proportions.

I have described my reaction to this catastrophic event several
times to my friends in the USA (I moved to the University of
Rochester in Rochester, New York, in 1963 and have been a citizen
of the USA since 1969) and they react to it with some surprise.
What I thought to myself was, 'Thank God, now we will win the
war—because the USA will be forced to enter the fray'. I hoped I
would play some role in the war and I did, in the Royal Canadian
Navy which I joined shortly after receiving my BSc in engineering
physics from Queen's in the spring of 1944. There was never any
doubt in my mind, after the horror of Pearl Harbor and the USA's
declaration of war on both Japan and Germany, what the eventual
outcome would be.

While I was serving in the Canadian Navy word came on 12
April 1945 that President Roosevelt had died in Warm Springs,
Georgia. I am sure there were few Canadians or others in the
allied countries battling against Hitler's waning might who did
not quietly weep at the loss of the leader of the country that,
despite its late entry into the war, was playing the major role in
defeating the Nazis. Harry S Truman became President of the
United States of America. Germany finally surrendered on 7 May
1945.

On 6 August 1945, on the orders of President Truman, an atom bomb with a yield equivalent to 12 500 tons of TNT exploded at a height of 1900 feet (580 metres) above Shima Hospital, almost the exact centre of the Japanese city of Hiroshima. It was dropped from a US Air Force B-29 bomber, the *Enola Gay*, piloted by Lieutenant Colonel Paul Tibbet, a 29-year-old 'veteran' flyer. By the end of 1945 the death toll at Hiroshima due to this cataclysmic event was estimated at 140 000. Three days later a second atom bomb was dropped on Nagasaki from another B-29 US Air Force bomber. It exploded at a height of 1650 feet (503 metres) with an estimated force of 22 000 tons of TNT. By the end of 1945 70 000 citizens of that city were dead. Most of these relatively quick deaths in the two cities were a result of blast and heat from the explosion. On 15 August 1945 the Emperor of Japan announced the unconditional surrender of his country. For the Allies the successful end of World War II, presaged by Japan's infamous attack on Pearl Harbor, had come at last.

Wrenching descriptions of the bombing of Hiroshima by survivors of the event can be found in the books *Hiroshima* by John Hersey [1] and *The Making of the Atomic Bomb* by Richard Rhodes [2]. Hersey's account first appeared as an article in a single issue of the *New Yorker* magazine on 31 August 1946 and later as a book published by Alfred A Knopf, New York. It is the story of six people who survived the explosion. Rhodes' book, published by Simon and Schuster, is one of the recent great pieces of scientific writing. It is a history of the genesis and development of nuclear physics leading to the manufacture and use of the atomic bomb. In the penultimate chapter, 'Tongues of Fire', Rhodes quotes many of the survivors of Hiroshima who describe their experiences. They speak of the unspeakable. Almost half a century has passed. A third atomic bomb has yet to reign terror on any human population. Those who believe in the efficacy of prayer, should pray that it never will.

I have often wondered, through the many years that have elapsed since these momentous events, what Tibbet felt as he gave the order to release his hellish cargo on the helpless people of Hiroshima who were mainly civilians. Did he really comprehend the awesome power of the weapon in the bomb bay of the *Enola Gay*? He certainly knew of the bomb test at Alamogordo but had not been there to view it. He and the crew of the *Enola Gay* had been thoroughly briefed on what precautions to take to avoid exposure to the light, heat and blast effects of the explosion. If he had known what destruction and death was to be visited on Hiroshima he probably salved his conscience in the belief that he was saving many US soldiers, sailors and marines from death in a more conventional battle that would have to be waged if it was necessary to invade Japan. Some predictions of the death toll of American forces, in such an event, gave estimates of over a quarter of a million, and of Japanese soldiers and civilians even higher.

I have wondered, even more, what the pilot of the second B-29, who dropped the atom bomb on Nagasaki a mere three days later, felt. The pilot of that plane was Major Charles W Sweeney, although it was under the command of Navy Commander Frederick L Ashworth. Did they question whether there had been time enough for the Japanese authorities to react to the bombing of Hiroshima to start suing for peace? Did it need 70 000 more deaths to bring Japan to her knees? I suppose the US leaders argued the second atom bomb would lead the Japanese to suppose the USA had a plethora of such weapons that they would continue to use in ever-increasing numbers on Japanese cities. Those two atom bombs and the one tested at Alamogordo were, in fact, the only three the USA had built to that time. No more were needed. However, could not the first have been the last and done the job as well? Wisdom in hindsight is wonderful but, of course, we can never know the answer. I still feel, even after all these years, that the dropping of the second bomb should have been delayed at least long enough to determine whether the devastation of Hiroshima had been sufficient to convince the Japanese authorities to sue for peace.

My undergraduate training at Queen's University, although it included a modicum of nuclear physics, naturally did not cover the then very new field of atomic energy so nuclear weapons were a complete surprise to me as they were to the vast majority of people in the world. Remarkably, then, when I was discharged from the Canadian Navy where I had risen to the exalted rank of lieutenant (equivalent to a captain in the army) it was with the proviso that I report for a position in Canada's atomic energy establishment at Chalk River, Ontario, a laboratory of which I had no previous inkling. My wife and I arrived in Deep River, the remote northern townsite for the atomic laboratory, in the depths of the winter of 1945. It resembled what I imagined a town in Siberia would be like. We lived in a four-room wartime house without a basement and heated with a wood-burning pot-bellied stove called a Quebec heater. The temperature often fell to 40 below zero outside (the same on both the Fahrenheit and Celsius scales) and when I wasn't working I seemed to spend most of my time chopping wood.

Because the house had no basement the floor was uncomfortably cold. Heat from the wood stove rose to the ceiling so the temperature gradient between floor and ceiling was 50 °F or so. Our first Christmas tree, in this hostile environment, after a week or so graded in colour from green at the bottom to brown at top. After enough snow had fallen I could bank it against the walls of the house as insulation, and that helped considerably.

At work my time was mostly spent in piling large blocks of pure graphite in various configurations around a lattice of uranium fuel elements as part of a low-energy nuclear reactor or 'pile' as it was called in those days. Each evening when I returned home to chop wood I was black to the skin from graphite dust. That, and my wood-hewing labours, determined me to leave what is now called Atomic Energy of Canada, Ltd (AECL) to acquire a PhD in nuclear physics. Apart from getting a more congenial job when I returned to AECL, I knew that, armed with a PhD, I would be entitled to a much better house in Deep River—at least one with a furnace. Incentives to pursue a higher education take many forms! I entered MIT in the fall of 1946 and was awarded my PhD in 1950.

I stayed on at MIT as a research associate in the Department of Physics until 1952. At that time I decided to return to Canada. I was appointed as an associate research officer in the Nuclear Physics Branch of AECL in Chalk River, this time doing nuclear physics research using a conventional single-ended Van de Graaff accelerator. As mentioned in Chapter 4, it was during the ten years following 1952 that the first double-ended or tandem Van de Graaff was purchased by AECL to replace the one I and my colleagues were using. One of the reasons I decided to return to Canada was the advent of what came to be called McCarthyism in the USA. Although I had probably nothing to fear from Senator McCarthy directly, he was such a repugnant person and his tactics of smearing people with the charge of Communism were so vile that I was finding it difficult to live in a country that would permit such outrages.

It was quite soon after my return to Chalk River in 1952, that the first and only nuclear reactor accident with which I was involved, albeit peripherally, occurred. While I was at MIT, a research reactor, called NRX, had been constructed at the AECL 'plant', as the laboratory at Chalk River was commonly called. It was not used for producing electrical energy, but it had many other uses in fundamental and applied research because of the very high flux of thermal neutrons it generated. As mentioned below it also produced plutonium-239. At the time of the incident in question, reactor physicists at Chalk River were subjecting NRX to some sort of subcriticality tests and had altered the operations of its control system.

Without any constraints, the power of a reactor can rise to dangerously high levels to the point where the heat produced by the fission process is sufficient to cause melting of the metal cladding of the uranium fuel elements and worse. The constraints are provided by cadmium control rods that are inserted into the reactor's core. Cadmium has a very high cross section for absorbing thermal neutrons and hence of slowing down the rate of fission and thus lowering the reactor's power and the temperature of the fuel elements. Normally these control rods respond automatically to changes in power and keep the reactor operating at a steady and safe power level. During the tests that were being made that fateful day the control rods were being operated manually.

Suddenly something occurred during the tests that caused the reactor power to rise precipitously. Before the scientists operating the control rods realized what was happening, and before they could insert these rods to shut the reactor down, radioactive leaks had developed in both the fuel rods and in rods containing cobalt that were used to supply cobalt-60 for medical uses. The reactor shut-down was, finally, successfully completed but only after considerable radioactivity had been released inside the reactor building. Most of us, not at the scene of the accident and unaware of what had happened, were surprised and alarmed when the sirens throughout the plant gave out the coded message. The decision had been made to evacuate the entire Chalk River laboratory, except for emergency personnel.

During this hasty evacuation there were a few minor automobile accidents on the five mile drive from the plant to the main highway that led to the townsite of Deep River. Most of us made the journey on buses provided by AECL and so suffered no mishaps. Part way along the road leading from the plant to the highway an emergency radiation monitoring unit had been quickly organized and after we had been checked for radioactivity we finally arrived home safely. We soon learned that officials of

AECL had euphemistically dubbed the incident to be the result of a 'pin hole' leak in fuel rod cladding. It turned out to be a little more than that.

A major clean-up of the radioactivity spread throughout the reactor building had to be undertaken. A pipeline was laid from the building to part of the deserted forest area surrounding the plant to get rid of contaminated cooling water. Units of the US Navy were brought in to remove concrete from all the interior walls of the building to a depth that reached uncontaminated concrete. This was deemed to be a part of their training as sailors who would be manning nuclear powered submarines. Jimmy Carter, then a member of the USA's nuclear navy and later President of the United States, was part of the team. A number of tales, some of them doubtless apocryphal, were told of incidents that took place during the clean-up.

One of these concerned some garbage cans located directly under the reactor vessel, called the calandria, that filled up with radioactive water that leaked from the calandria. Remotely controlled TV cameras surveyed the scene and high-level conferences were held to determine how to safely remove them. One day it was discovered that the cans were miraculously empty. It was finally determined that some unsuspecting janitor had entered the room and dumped them down the drain.

Another story concerned the rods filled with pellets of cobalt that had been removed and stored in underground tanks filled with water. Some of the highly radioactive pellets had fallen out, presumably through the 'pin holes', and were lying in the bottom of the tanks. Some wit in the radiobiology section of AECL suggested the following ingenious solution. It is known that turtles, particularly snapping turtles, were more impervious to nuclear radiation damage than most other living creatures. Cobalt has magnetic properties. Why not fasten magnets to the underside of the turtles' shells, attach them to a long string, lower them into the tanks and let them crawl slowly around while their magnets harvested the loose cobalt pellets? Why not indeed. So far as I know the plan was never even initiated, much less carried out.

Eventually the clean-up of the reactor building was effected. After that, and with great difficulty, the calandria was removed and transported in the dead of night on a flat bed truck out of the plant to a burial side in the surrounding northern woods. The driver of the truck was shielded by a thick lead wall from the

*Photograph of the 'Little Boy' bomb used on Hiroshima. The bomb used a gun barrel within which a uranium-235 bullet was fired against a uranium-235 target. It was 10 feet long and 28 inches in diameter and weighed almost 9000 pounds. It was the only weapon of such construction and, consequently, had never been tested. The bomb dropped on Nagasaki three days later, called 'Fat Man', was an implosive device employing plutonium-239. A weapon like it had been tested at Alamogordo, New Mexico on 16 July 1945.*

highly radioactive calandria. NRX was then completely rebuilt and is still operating to this day.

The main purpose of AECL was not to build atomic·bombs (Canada has never done so—why should they when their friendly southern neighbour had a plentiful supply?) but to harness atomic energy to provide electric power for homes and industries. However, a by-product of the NRX reactor at AECL's Chalk River laboratory was an isotope of plutonium, plutonium-239. It was sold to the USA for use in nuclear power reactors and atomic bombs.

The bombs that levelled Hiroshima and Nagasaki were quite different from each other. 'Little Boy', as the Hiroshima bomb was dubbed, employed a gun barrel in which a 'bullet' of the rare isotope of uranium, uranium-235, was fired against a uranium-235 target. The fusion of these two sub-critical masses of uranium-235 produced a critical mass that caused the uranium nuclei to undergo rapid fission. The instantaneous energy released in this nuclear process produced a detonation in which the gun barrel was vaporized. The heat and explosive shock wave that followed was responsible for most of the deaths that occurred in the first

"An atomic bomb," the Japanese study of Hiroshima and Nagasaki emphasizes, "... is a weapon of mass slaughter." A nuclear weapon is in fact a total-death machine, compact and efficient, as a simple graph prepared from Hiroshima statistics demonstrates:

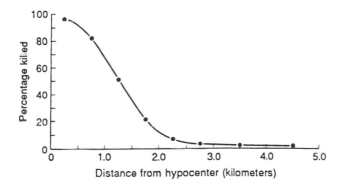

*Percentage of the people killed immediately at Hiroshima as a function of their distance from the hypocentre (the point on the ground directly below the explosion). The bomb was exploded at a height of 580 metres above the ground, so zero on the distance scale shown here corresponds to 580 metres. From [2].*

few days. Those who did not die during this initial period, however, seemed at first to improve but then to sicken again. Now it was radiation-induced illness caused, it was believed, primarily by gamma radiation produced in the nuclear reactions that accompanied the fission process that was the culprit. From whatever the cause, fewer than 10% of those people within half a kilometre (a third of a mile) of the hypocentre (the point directly below the explosion) managed to survive. The percentage of survivors increased rapidly, reaching 80 to 90% at distances of 2 kilometres.

By the mid 1980s there were approximately 60 000 Japanese citizens who had been exposed to and survived the radiation from the Hiroshima bomb and some 30 000 in the case of Nagasaki. These survivors are the best studied human beings in history in terms of the short- and long-term effects of nuclear radiation. For such studies to be meaningful, however, one must have a good estimate of the nature and degree of the nuclear radiation they received. Little Boy was a 'one of a kind' atom bomb and so had never been tested. No one knew with any accuracy at all what its neutron and gamma ray yield might have been, although some relatively crude computer calculations carried out a few years after the explosion indicated that neutrons were the primary culprit.

In the early 1960s, in an attempt to answer the question, an unshielded reactor vessel was erected on top of a 1600 foot high steel tower in the Nevada desert. Neutrons from the fission of its nuclear fuel radiated in all directions from the reactor and their flux was measured as a function of distance. To make the measurements even more realistic a village was built with structures mimicking those of a Japanese village so the shielding effects of such structures against neutrons could also be evaluated. Not too surprisingly, the results of this expensive project, called the Bare Reactor Experiment Nevada, were ultimately called into question on the grounds that the reactor's neutron energy spectrum probably differed substantially from that of Little Boy in which a massive amount of gun barrel iron was present.

In the 1980s it was decided to construct an exact replica of Little Boy and to conduct measurements of the neutron energy spectrum it emitted when two uranium-235 masses were brought together close enough to produce neutrons but not so close as to cause an explosion. It was an exceedingly delicate experiment, carried out at Los Alamos National Laboratory in New Mexico. These tests indicated that it was gamma rays and not neutrons that were responsible for the radiation doses to the Hiroshima survivors. Again, however, the extent to which this experiment mimicked the actual results of the explosion in which the gun barrel was vaporized was in doubt.

It was finally decided that only by complex computer modelling could the character and the intensity of the radiation resulting from Little Boy's explosion be reliably estimated as a function of the the slant range (the distance of points on the land from the centre of the explosion 580 metres above the ground).

Such computer modelling was carried out in 1986. It took account of the massive iron casing of the gun barrel, as well as meteorological, topographical and other factors. The gamma ray intensity emitted in the explosion could be measured as a function of distance from the explosion by a technique called thermoluminescence dosimetry. These measurements were made in roof tiles some omniscient Japanese scientists had collected and stored, carefully cataloguing them according to their distance from the bomb's hypocentre and the extent to which they were shielded from the bomb by nearby buildings. The computer calculations were in general agreement with the measured gamma ray intensities. These calculations showed that neutrons played virtually no role at all. The Hiroshima bomb's lethal radiation was mainly gamma rays—so said the computer.

Both neutrons and gamma rays can produce deleterious biological changes to human cells, and so it was troublesome that the radiation damage to the survivors of the Hiroshima bomb was, apparently, due to gamma rays. Exposure of people to sources of gamma rays and x-rays (both forms of electromagnetic radiation) is much more common than exposure to neutrons. X-rays and gamma rays are widely used in medical diagnostics and in the treatment of cancer. Exposure to neutrons, on the other hand, occurs mainly to people associated with nuclear reactors or scientists using accelerators for nuclear or particle physics research. Both groups comprise a relatively small fraction of the population. If the radiation damage suffered by survivors of Hiroshima was mostly gamma ray induced this suggested that one would have to be much more conservative about the dosage of gamma ray and x-ray radiation people should be permitted to receive.

In the mid to late 1980s and early 1990s some measurements [3] of the neutron intensity produced by the Hiroshima bomb were made that indicated it was much higher than the computer calculations predicted, most notably at slant ranges beyond 1 kilometre. These involved measurements of the production of radioactive isotopes of cobalt and europium produced when the appropriate stable isotopes of these elements capture the bomb's neutrons. The problem is that these radioactive isotopes have half-lives (the time it takes for half of them to decay) that are short compared with the some 45 years that had elapsed since the bomb exploded.

The bomb that exploded over Nagasaki, dubbed 'Fat Man', on the other hand, was of a different design. It was an implosive device that employed plutonium-239. The latter had been produced in large reactors in the USA and at Chalk River, Canada. In such a bomb a sphere of plutonium was compressed uniformly with tremendous force by an outer mantle of conventional high explosives. A weapon like this had been tested at Alamogordo, New Mexico on 16 July 1945 and both its gamma ray and neutron production intensities had been measured. The computer modelling gave much better agreement for these intensities for the Nagasaki bomb than it appeared to be giving for the one dropped on Hiroshima—at least for the latter when it came to neutrons.

The activities of the cobalt and europium samples, collected in building material at various distances, were measured by decay counting, that is by measuring the radiation emitted as the radioactive atoms produced by the bomb decayed to stable atoms. Most of the present-day survivors of the Hiroshima bombing were at slant ranges of 1 to 2 kilometres from the blast centre. Beyond about 1.5 kilometres there is not enough radioactive cobalt and europium left to measure. What one needs to find is a radioactive element with a half-life that is long compared with the time that has elapsed since the bomb went off and that the bomb's neutrons could have produced with high probability at these large distances. Furthermore it should be produced by neutron capture by some relatively abundant stable element present in cement, granite and clay tiles that were located at such distances from the blast epicentre.

In the late 1980s it was realized by scientists at the Lawrence Livermore National Laboratory in California and at the Technical University in Munich, Germany that a certain isotope of chlorine was ideal. That isotope was chlorine-36.

Here I remind the reader what a nuclear isotope is. The nucleus of any atomic element is made up of positively charged particles called protons and neutral particles called neutrons (particles of equal mass to the proton but having no electric charge) packed closely together and held together by the nuclear force. The number of positively charged protons in the nucleus determines the number of negatively charged electrons that can be bound in orbits around the nucleus. It is the number of electrons that determines the chemical properties of the atom. Chlorine has 17 protons in its nucleus and thus 17 electrons circling around it. There are, however, two stable isotopes of chlorine with 18 and 20 neutrons respectively in their atomic core or nucleus. So one isotope contains 17 protons and 18 neutrons, or 35 nucleons and the other 17 protons and 20 neutrons, or 37 nucleons. These two isotopes, chlorine-35 and chlorine-37, as mentioned above, are stable. For reasons fathomable to nuclear physicists, if the nucleus of a chlorine atom contains 19 neutrons, i.e. chlorine-36, it is no longer stable but decays to stable argon-36 with a half-life of 301 000 years by emitting an electron from its nucleus.

If stable chlorine-35 finds itself in a flux of neutrons, such as produced when an atomic bomb explodes, some of the nuclei capture a neutron to form chlorine-36. Chlorine is reasonably ubiquitous and can be found, at low concentrations, in building materials in Hiroshima that survived the nuclear holocaust of that fateful day in 1945. Because chlorine-36, although unstable, has such a long half-life, whatever amount of it was produced on 6 August 1945 is still there to this day. The questions facing the scientists at Munich and Livermore were, firstly, are there samples of concrete, roof tiles and granite, perhaps from graveyard tombs, whose locations from the centre of the explosion is accurately known, and can they be obtained? And secondly, if they can, would it be possible to detect quantitatively the exquisitely small amounts of chlorine-36 in the material? The answer to the latter question is no, if one attempts to measure the electrons emitted when the chlorine-36 decays. It would take many years by such direct decay counting—just because chlorine-36 decays at such a slow rate due to its long half-life.

How long, for example, would it take, by decay counting, to make a measurement with 10% accuracy of the chlorine-36 to stable chlorine ratio in a 1 milligram sample of chlorine in which the ratio is $10^{-11}$ (the value actually measured at 1 kilometre)? The answer is 4 months. At 1.5 kilometres, it turns out, there is a further factor of 20 decrease in the ratio so it would take 7 years! Quite clearly decay counting was out of the question.

As recounted in Chapters 2 and 3, accelerator mass spectrometry (AMS) had been developed at the University of Rochester beginning in 1977. AMS measures the radioactive and stable isotopes directly without the need to detect the radioactive decay of the former. It had great sensitivity, high accuracy and used milligram sized samples. It is just what was needed to make the measurements of the ratio of chlorine-36 to stable chlorine in building materials from Hiroshima. Measurements at Rochester had already shown that AMS could measure these chlorine ratios to about two parts in $10^{15}$ using samples of AgCl of 1 milligram or less.

However, was the appropriate building material, whose distance from the explosion centre was accurately known, actually available after all these years? It turned out that some of it could be collected from the remains of concrete structures still in their original location and, in one case, from a granite tombstone located 590 metres from the centre of the explosion. In addition, due to the incredible foresight of a few Japanese scientists, it was possible to obtain roof tiles. As mentioned above, these had been collected at some time after the explosion, their exact location had been carefully recorded and they had been then stored for possible future use. They had already been used to measure the

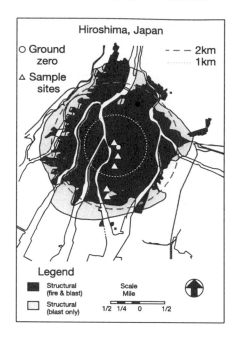

*A map of Hiroshima, Japan showing the approximate location of ground zero (the hypocentre) and the locations of the five sample sites where chlorine-36 was measured.*

intensity of gamma rays emitted from the nuclear explosions by the technique of thermoluminescence. Building materials like this could be obtained at distances out to 2 kilometres from the centre of the explosion and from them the chlorine could be extracted— enough to measure the chlorine-36 to stable chlorine ratio by AMS. This ratio is readily converted to slow neutron flux density (the number of slow neutrons per square centimetre) from the bomb.

A collaboration between scientists at Hiroshima University and the Technical University of Munich, using the Munich AMS facility, resulted in the measurement of the chlorine-36 to chlorine ratio (and thus the slow neutron fluence) in a granite gravestone 107 metres from the hypocentre (slant range of 590 metres) of the Hiroshima explosion [4]. Although the measurement is interesting it gives no clues about the effect of such slow neutron doses on humans since everyone at that distance from the explosion died instantly from blast and heat.

It was a scientist at the Lawrence Livermore National Laboratory (LLNL) in California, Tore Straume, who reasoned that AMS could be employed to make chlorine-36 to chlorine ratio measurements in building materials at Hiroshima at much greater distances from the hypocentre where there actually were survivors. Straume is a biophysicist whose work at Livermore over the past 18 years has been in the field of nuclear radiation research.

American and Japanese governments had funded, during the last 40 years or so, costly studies of the effects of the radiation from the bomb on these survivors. It was of crucial importance to know whether the radiation was primarily neutrons or gamma rays or a mixture of both. The elaborate computer calculations carried out in 1986 said it was mainly gamma rays. Their flux density had already been measured in roof tiles and measurement and calculation agreed. Straume felt the cobalt and europium neutron activation data were indicating that the calculations underestimated the slow neutron flux density. Some time in 1988 he suggested to the AMS group at Rochester this be verified by measurements of chlorine-36 and that these measurements be made out to distances as great as possible. At that time Rochester was the only AMS laboratory in the USA that had the expertise and instrumentation to measure chlorine-36. We enthusiastically agreed to collaborate with him on the project.

What established the maximum distance to which the measurements could be made was the background of chlorine-36 found in the outer surface of rocks, other geological samples and concrete building material that had never been exposed to neutrons of anthropogenic origin. Such chlorine-36 is produced by cosmic ray neutrons that make it through the earth's atmosphere and get captured by stable chlorine-35 in the rocks. The chlorine-36 to chlorine ratio from this cosmogenic source was measured at Rochester in Hiroshima concrete that had been totally shielded from bomb neutrons. At about a slant distance of 2 kilometres this background level was reached making the measurement of bomb-produced neutrons impossible at greater slant ranges.

We measured five samples collected from Hiroshima at slant ranges varying from 650 to 1700 metres. When the results were translated into thermal neutron flux densities they showed that a person at the 650 metre distance would have received a blast of neutrons in the split second of the explosion equal to the thermal neutron dose permitted for human beings to receive in a period of over 500 years. Even at the largest distance of almost 2 kilometres the neutron flux density was equivalent to a permitted thermal dose of 2 years. The latter was almost a factor of 100 greater than the computer calculations had indicated. It was almost certainly

### NEUTRON DISCREPANCIES IN THE DS86 HIROSHIMA
### DOSIMETRY SYSTEM

T. Straume,* S. D. Egbert,† W. A. Woolson,† R. C. Finkel.* P. W. Kubik,‡ H. E. Gove,‡
P. Sharma,‡ and M. Hoshi§

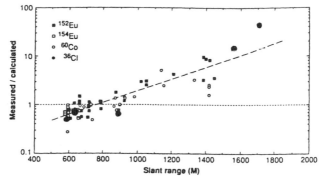

Fig. 1. Ratios of measured to calculated neutron activation in Hiroshima at various distances from the epicenter (slant range). Measurement results for $^{152}$Eu (■), $^{154}$Eu (□), and $^{60}$Co (O) were obtained from previously published reports and papers (Hashizume et al. 1967; Hashizume 1983; Nakanishi et al. 1983, 1987; Hasai et al. 1987; Maruyama and Kawamura 1987; RERF 1987a,b; Sakanoue et al. 1987; Hoshi et al. 1989; Kerr et al. 1990; Kimura et al. 1990). Measurement results for $^{36}$Cl (●) are from Table 1. Detailed Monte Carlo modeling calculations for each sample using DS86 spectra and fluences were performed here as described in the text. The dashed line is the least-squares best fit to all the data points.

| RADIOISOTOPE | HALF LIFE (YEARS) |
|---|---|
| $^{36}$Cl | 301,000 |
| $^{60}$Co | 5.3 |
| $^{152}$Eu | 13.2 |
| $^{154}$Eu | 8.5 |

*Ratios of measured to calculated neutron activations at various slant ranges from the Hiroshima bomb explosion. The chlorine-36 production was through the thermal neutron capture $^{35}$Cl(n, $\gamma$)$^{36}$Cl. The DS86 calculation with which the data are compared took account of the massive steel gun casing and of meteorological and topographical factors existing at Hiroshima. The broken line is a least squares fit to all the data. A fit to just the chlorine-36 data would have yielded even greater discrepancies with the DS86 calculations at distances beyond 1500 metres. Because of its much longer half-life only chlorine-36 can be measured reliably beyond this distance. See [7].*

neutrons that had made the major contribution to the radiation damage suffered by the Hiroshima survivors, after all, and not gamma rays. This meant that the exposure of humans to x-rays and gamma rays was not as dangerous as one would have to assume if the Hiroshima bomb's radiation was mainly gamma rays.

These data were of the greatest significance. Straume presented a preliminary version of the results of this Livermore–Rochester collaboration at the Fifth International Conference on Accelerator Mass Spectrometry that was held in Paris, France in April 1990 [5].

**Table 6.1.** *The $^{36}Cl/Cl$ ratio translates directly to thermal neutron flux density (neutrons/cm$^2$) via the chlorine-35 neutron capture cross section of 44 barns ($10^{-24}$ cm$^2$). The table gives the flux densities as a function of slant range as well as their translation into permitted thermal neutron dose in years. At the two largest distances people received almost twice a year's worth of the permitted dose during the brief explosive flash.*

| Slant range (metres) | $^{36}Cl/Cl$[a] ($\times 10^{12}$) | Neutrons/cm$^2$ ($\times 10^{-9}$) | Permitted thermal neutron dose (years) |
|---|---|---|---|
| 590 | 113.0[b] | 3420 (445) | 685 |
| 653 | 89.8 | 2720 (109) | 545 |
| 891 | 6.37 | 192 (12) | 38 |
| 1562 | 0.304 | 9.2 (1) | 1.8 |
| 1708 | 0.327 | 9.9 (3) | 2 |

[a] Background subtracted.
[b] Kato *et al* 1988 *J. Radiat. Res.* **29** 261.
The thermal neutron induced reaction $^{35}Cl/(n, \gamma)^{36}Cl$ has a cross section of 43.6 barns and this value was used to convert $^{36}Cl/^{35}Cl$ ratios to neutrons/cm$^2$.
The maximum permissible slow neutron dose is a 5 rem/year and that corresponds to a neutron fluence of $5 \times 10^9$ neutrons/cm$^2$ per year.

Two years later he presented a final version of our data at the big annual meeting of the American Chemical Society that was held in San Francisco [6]. By this time another AMS laboratory had entered the business of measuring chlorine-36. It was located in Straume's home laboratory, Lawrence Livermore.

The man who headed that AMS laboratory was a well known physicist, Jay Davis, whose latest claim to fame, at that time, was as one of the international inspectors who had gone to Iraq to determine how far along they had been in producing a nuclear weapon at the time they invaded Kuwait. Davis is a very outgoing person with a strong personality and a tremendous flair for publicity. He chaired the session at which Straume gave his paper. There was one new measurement of the chlorine-36 to chlorine ratio in chlorine extracted from a concrete sample from Nagasaki that Tore presented at this meeting. The measurement had recently been made at Livermore and was in agreement with the 1986 computer calculations. I was mildly annoyed at this because I thought it should have been measured at Rochester— but Livermore was Straume's home laboratory after all and it was really up to him to decide where he wanted the measurements to be made. Besides, I am sure Jay put a lot of pressure on Straume to use the Livermore AMS facility and Davis is a man who is hard to resist. The AMS laboratory at Livermore, at that time was particularly noted for the pioneering work it was doing in biomedicine. However, at the end of Straume's talk, Davis made the assertion that, if the Livermore AMS laboratory had done nothing else than make that single measurement of chlorine-36 in Nagasaki concrete, the considerable expense (to the US taxpayer) of creating the AMS laboratory at Livermore would have been fully justified. I responded by telling Jay how pleased we at Rochester had been to assist Livermore by making all the chlorine-36 measurements in material from Hiroshima except the one near the hypocentre measured at Munich.

As noted above, the Hiroshima measurements were very significant and I knew they would engender considerable valuable publicity for both the AMS group at the University of Rochester's Nuclear Structure Research Laboratory and for Straume and his collaborators at Lawrence Livermore. Tore and I discussed where they should be published. I suggested the journal *Science*, probably the most prestigious general science journal in the USA. As recounted previously, it was that journal in which we had published our seminal paper on the invention of AMS and its first application to the measurement of carbon-14 in natural organic material in 1977. This led to the use of AMS in carbon dating that has now virtually supplanted the old technique of decay counting that won the Nobel Prize for chemistry for its inventor Willard Libby in 1960. Publication in *Science* would ensure that the article would come to the attention of the widest body of interested scientists and to the public at large. Tore agreed with that suggestion. He then proceeded to prepare an article on the results and submitted it to *Science*. *Science* is published weekly and each issue contains only one or two articles. The editors informed Straume that they had a long list of articles lined up for publication and so they were unable to accept his. When he informed me of this I tried to convince him to condense it to a report length paper and resubmit it to *Science*. Reports are two or three pages long and a dozen or so are published each week in *Science*. He said he could not do the material justice in a paper of such a short length despite the fact that, as I pointed out to him, he could write a longer version later and publish it somewhere else. He said he had decided to submit it to *Health Physics*. I was disappointed but, since he was the senior author, he could call the shots. It was published in that journal in October 1992 [7].

Because it is valuable for the University of Rochester to get publicity for important scientific breakthroughs that take place in its laboratories I responded sometime later to a query from the science writer of the university's public relations office, Tom Rickey, asking whether our AMS group had been doing anything noteworthy lately. I mentioned our Hiroshima measurements and said they were about to be published. I noted that Tore Straume of Livermore was the senior author and that I would discuss with him the possibility of a joint Livermore–Rochester press release. I did so and Rickey also talked to him. We both had the impression that Straume had agreed to this suggestion.

Not long after that, I received a copy of a press release dated 1 October 1992, issued unilaterally by Livermore concerning the Hiroshima chlorine-36 measurements. It gave short shrift to the role our AMS group had played in the work. In a note attached to the Livermore press release that Tore faxed me he wrote 'In discussions with your press person we agreed that it would be most appropriate for LLNL to do the release and send you a copy. As you see we have given proper credit to Rochester.' That was not the agreement that Rickey and I thought we had with Straume and, although the Rochester AMS group was mentioned, more or less in passing, the release ensured that subsequent publicity would involve Livermore almost exclusively, as it did.

The press release was picked up by the *New York Times* and an article based on it appeared in the *Science Times* section of that paper on Tuesday, 13 October 1992 written by one of their top science writers, William J Broad. It was entitled 'New study questions Hiroshima radiation' and was an extensive and exceedingly well written article. It used the press release as a springboard to include much additional relevant information based on research Broad carried out and on interviews he conducted with many scientists including Straume but no one from Rochester. The only mention, in the article, of the method used to make the chlorine-36 measurements, a technique that was crucial to the success of the project, was 'The main tool was an accelerator mass spectrometer, a device perfected in the 1980s, that could measure in great detail the subtle differences between chemical isotopes.' The names of all the authors on the paper and their places of employment, including me and my colleague in our AMS group, Pankaj Sharma, were listed toward the end of the article and that was the only mention of the University of Rochester. Both Tom Rickey and I felt very badly about this but there was not much we could do. Rickey managed to get a more balanced account published in one of the local Rochester newspapers but that was a far cry from the *New York Times*. I was disappointed in the way Straume had handled the whole affair. The Lawrence Livermore National Laboratory has had a somewhat checkered history, partly because much of the research carried out there is shrouded under the cloak of excess secrecy and partly because of the contentious personality of its founder, Edward Teller. Maybe some of this tended to rub off on some of its scientists.

More recently Straume [8] has had additional measurements of the chlorine-36 to stable chlorine ratio made at the Livermore AMS laboratory on Hiroshima samples. These samples were collected at slant ranges between the two Rochester measurements made at about 900 and 1600 metres. This is a critical range of distances since it corresponds to survival rates of about 20 to 60%. The results agree with the original Rochester data and now present a reasonably complete picture of the slow neutron fluence emitted by the Hiroshima atom bomb over distances at which there were substantial numbers of survivors. What remains to be done is to evaluate the new information on the radiation these Hiroshima survivors actually received and to compare the doses received with the biological damage that resulted. It seems remarkable that it took almost 50 years for this to happen, and it would not have happened yet but for the invention of accelerator mass spectrometry in 1977. Tore Straume and his Livermore collaborators deserve much, but not all, of the credit for these important Hiroshima revelations.

# Chapter 7

# The Initial Peopling of the Americas

When did the first humans reach the Americas and who were they? The answers to these questions are still controversial and unsettled. Radiocarbon dating, especially the use of accelerator mass spectrometry (AMS) for the detection of carbon-14, is beginning to provide some answers. It has already settled the question of whether the Vikings discovered America before Columbus—there seems little doubt they beat him by at least 500 years. Although the Vikings were arguably also the first Europeans to reach America it had clearly already been discovered by other people many millennia before. The Vikings, after all, in their explorations of North America, encountered native Americans, both Indian and Inuit. When Columbus arrived in America he thought he had reached Asia, in part because the Indian inhabitants looked Asian to him. Where and when did these native people originate? Of one thing we can be sure; their origins and first arrival, at least in North America, must have been affected by the periodic glaciations of the North American continent.

The earth has been subjected to many epidemics of glaciation throughout its history. The periodicity of such glaciations, although somewhat irregular, seems to be around 100 000 years. The last glacial maximum occurred some 21 000 years ago and the one prior to that about 70 000 years ago (an exception to the more normal 100 000 year period). These are now believed to be a result of temperature changes caused by three cycles of the earth/sun geometry that control the distribution of solar energy radiant on the surface of the earth. They are the tilt of the earth's axis of rotation with respect to the plane of its orbit around the sun, the eccentricity of the earth's orbit around the sun and the precession of the equinoxes. The first of these causes the seasonal variations in the northern and southern hemispheres. The second reflects the fact that the earth's orbit around the sun is not circular but slightly elliptical. On average the earth is 93 million miles from the sun but at its closest distance (perigee) it is 3 million miles closer to the sun than at its greatest distance (apogee). This alone can cause a 7% difference in the radiant energy the earth receives from the sun. The precession of the equinoxes is due to variations in the tilt of the earth's axis with respect to the plane of its solar orbit. Over the epochs this phenomenon causes different stars to become the 'Pole star'. It can also affect, to some extent, the distance of the northern and southern hemispheres of the earth from the sun. Since the average yearly temperature difference between a normal and a glacial climate is only about 10 °C (18 °F) there may be other factors that contribute to glaciation, in addition to the ones mentioned above, that are still not understood,

The maximum in the last great Ice Age in North America occurred some 21 000 years ago [1]. During the subsequent deglaciation, within a period of about 8000 years, a mass of ice equivalent to an increase by 100 metres in the global sea level was returned to the oceans as water. At its peak, a sheet of ice covered all of Canada, parts of the northern USA and all of Greenland. Surprisingly, even at the glacial maximum, substantial parts of Alaska were free of glacial ice. Between the longitudes of 75W and 90W glacial ice extended as far south as latitude 40N (mid Illinois, Indiana, Ohio and all of Pennsylvania). Between longitudes of 90W to 100W it went not quite as far south, to latitude 45N (mid South Dakota, Minnesota and Michigan). Further west, between 100W and 125W, it reached only to latitude 50N, approximately the present Canadian–USA border, covering the provinces of British Columbia, Alberta and Saskatchewan. There were two main ice sheets of separate origin. The one east of longitude 120W is referred to as Laurentide ice and that to the west covering the coastal regions of Canada and the USA is referred to as Cordilleran ice. The western edge of this ice sheet along British Columbia was available for human occupation as early as 14 500 years ago and may have served as a migration route that long ago. This conclusion is a result of the AMS carbon dating of organic material in marine cores from the continental shelf edge of British Columbia in the Queen Charlotte Islands [2]. By about 12 000 years ago, and perhaps somewhat earlier, a corridor had opened between the Laurentide and Cordilleran ice sheets providing an ice-free region from the Bering Straits in Alaska to northern Montana and points south. At the same time there may have been a land bridge across the Bering Straits connecting the Chukchi Peninsula in Siberia to the Seward Peninsula in Alaska. This land bridge submerged some 12 000 to 11 000 years ago and Alaska and Siberia became separated. By 7000 years ago most of North America was free of ice except for areas in northern Quebec and a patch on the northwest corner of Hudson's Bay, marking the onset of the Holocene (fully modern) epoch of earth history. Ice still covered almost all of Greenland, as it does today.

A continuing controversy among American archaeologists concerns the question of when humans first came to America. Did they arrive thousands of years before this last Ice Age or about the time it was ending? In the latter case a reasonable point of entry would have been from Siberia to Alaska. Today some 60 miles of water in the Bering Strait separates these two peninsulas. Even if no land bridge existed at the time the Laurentide and Cordilleran ice sheets parted, crossing such a relatively short stretch of water by boat would be entirely feasible. It is, however, most likely that these early humans, in pursuit of mammoths or other wild game, pursued their prey across a land bridge from Siberia to Alaska and began the population of North America.

Another possibility would have been for the earliest human visitors to America to have sailed down the western coasts of what are today Canada and the United States and to land and disembark at appropriate congenial landfalls. For that matter they could have continued their sea journeys all the way to South America. What is the archaeological evidence for any of these scenarios?

The oldest palaeoindian sites discovered so far in North America are located in Alaska, Montana, Wyoming, South Dakota, Colorado, Oklahoma, New Mexico and Arizona. Except for a site on Prince of Wales Island, discussed later, where bones of a human were found in a cave, none of similar antiquity have yet been discovered with absolute certainty west of longitude 115W, that is in the states of Washington, Oregon, Nevada or California. That is an argument against any landings from the sea on the west coast of North America south of British Columbia. It is considered more likely these sites east of longitude 115W were populated by people who had crossed into North America from Siberia. C Vance Haynes, Jr, a faculty member of the Departments of Anthropology

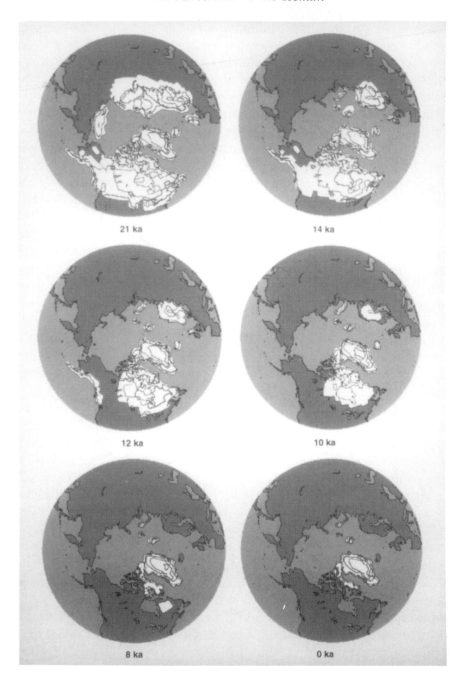

*The ice sheet coverage of North America during the last ice age at various times in the past. Some 13 to 12 thousand years (ka) ago an ice-free corridor had opened linking Alaska to northern Montana and points south. Courtesy of W R Peltier, University of Toronto.*

and Geosciences at the University of Arizona, Tucson, Arizona is one of the leading archaeologists/anthropologists in the USA. He, among others, has studied and analysed the evidence for the ages of these sites based on radiocarbon dating and concludes they are 11 000 to nearly 13 000 years old, with the oldest occurring in Alaska and the youngest in Arizona, Colorado, Montana, New Mexico, Oklahoma, South Dakota and Wyoming. In a 1992 paper published in a book dedicated to the fortieth anniversary of the first use of radiocarbon for dating by Nobel Laureate Willard Libby, Haynes discusses the radiocarbon dating of the peopling of the New World [3].

He notes that, in 1927, the small cattle shipping town of Folsom in northeastern New Mexico gained fame by the discovery, at a nearby site, of finely fashioned, fluted, stone projectile points associated with the bones of an extinct form of buffalo. The discovery of what became known as Folsom points here and in sites in six other states to the west, north and east of New Mexico and the radiocarbon dating of artefacts found in the sites clearly placed palaeoindians in North America at the ending of the glacial era some 12 000 years ago. This was about when the ice-free corridor between the Laurentide and Cordilleran ice sheets provided an opening from Alaska to the south. The 12 000 year age was a result of the most accurate dating of the Folsom site in New Mexico and it was provided by AMS. Haynes notes, in his article in the Libby retrospective, '... I did not anticipate the revolutionary technology of accelerator mass spectrometry (AMS). This ideal technique for dating single lumps or flecks of charcoal was applied to the type Folsom sample in 1987 by selecting five discrete lumps for individual analysis as well as a composite made by combining a small chip off each lump. The six analyses resulted in a statistical average of 10 890 ± 50 (radiocarbon years BP), the most precise radiocarbon date yet obtained for the Folsom complex.' This corresponds to a calibrated age of 11 990 years BP (before present) (referred to above as 12 000 years).

It should be explained that, as is customary in radiocarbon dating, ages are quoted in the number of radiocarbon years BP where present is taken as the year 1950 (before the major testing of atomic weapons, mainly by the USA, in the period between 1950 and 1960 injected a huge pulse of radiocarbon into the environment). Furthermore, following a quaint custom in radiocarbon dating that is used to this day, the half-life of the decay of carbon-14 is taken to be the best value known at the time to the father of carbon dating, Willard Libby, namely 5568 years and these are referred to as radiocarbon years. When one uses the more accurate modern value of 5730 years ages must be increased by 3%. A radiocarbon age of 10 890 years meant that the Folsom site was inhabited by *homo sapiens* some 11 200 years BP or in 9250 BC. However, this does not take into account the fact that the cosmic ray flux incident on the earth's atmosphere has varied with time. It is this flux that produces the atmospheric carbon-14. Calibrations for this variation are available, and when they are used the age of the Folsom site becomes more like 9950 BC or close to 12 000 years old. In what follows ages will be designated either as radiocarbon ages BP or calibrated ages BP.

In 1932 and 1933, fluted projectile points, resembling those of Folsom but more robust, associated with the bones of many mammoths were discovered in the railroad siding of Dent north of Denver, Colorado. Subsequently, similar projectile points were found near Clovis, New Mexico in stratigraphy underlying Folsom sites. Radiocarbon dating suggested that on average, however, Clovis man preceded Folsom man by only a few hundred years.

Haynes states in his article [3] in the Libby retrospective, 'Recent [by recent Haynes means 1989 and 1990] investigations in the Nenana Valley of central Alaska have demonstrated the presence there, at least as early as Clovis time, of a cultural complex that is Clovis-like in most respects except for the absence of the diagnostic Clovis point.' Radiocarbon dating of charcoal and soil organics at these Nenana sites gave calibrated ages ranging back 12 100 to 12 900 years BP. This prompted Haynes to suggest in 1963 and again in 1987 that the ice-free corridor allowed these Clovis progenitors to reach North America south of Canada. He speculates that they may have developed the technique of making fluted stone points on the way.

Additionally, and subsequent to the Haynes article cited above, palaeoindian-like flint projectile points were discovered in Alaska in conjunction with charcoal in shallowly buried hearths. The charcoal has been dated by AMS to be as old as 11 600 radiocarbon years BP. This corresponds to a calibrated age of 12 760 years BP. The results were reported by M L Kunz and R E Reanier in the 4 February 1994 edition of *Science* [4]. The artefacts were recovered from the Mesa site, a hunting lookout some 60 metres above the surrounding territory, on the northern flanks of Alaska's Brooks Range, at about latitude 67N and some three degrees north of the Nenana Complex. This strengthens the conviction that palaeoindians crossed into North America from Siberia to Alaska some 13 millennia ago and from there proceeded southward to Clovis and Folsom, New Mexico as well as to other sites in the northern and southern Central States and perhaps even further south to South America.

What is the evidence for early modern man in Siberia? It seems a remarkably inhospitable place for humans to elect to settle. In June 1995 American and Russian archaeologists reported two Siberian sites occupied by humans and dating to at least 40 000 radiocarbon years BP [5]. At one site, Makarovo-4, on a high bluff overlooking the Lena River, an accumulation of bone samples from woolly rhinoceroses, red and roe deer were uncovered as well as stone blades, knives and scrapers. It was some of these bones, radiocarbon dated at the AMS facility at the University of Arizona, that determined the age of the site to be at least 40 000 years. As will be discussed below, the use of AMS renders the dating of old bones more reliable than in the decay counting method of Libby with its need for large samples. It turned out, however, there was not sufficient similarity between the appearance of these implements and those of palaeoindian tradition to confidently relate the people inhabiting these Siberian sites near the Lena River to those at the palaeoindian sites in Alaska at Mesa. The age of the Siberian sites pre-dates those in Alaska by some 30 000 years—during which time, however, at the very least, it would have been possible for people to have migrated from the Siberian to the Alaskan sites.

In 1996 the discovery of the Uptar site near the city of Magadan in northeastern Siberia was reported in *Science* on 2 August 1996 by Maureen King, a University of Washington archaeologist, and S B Slobodin, a member of the Department of Education in Magadan [6]. The site contained a number of chipped stone artefacts, including one bifacial fluted point resembling those found at the palaeoindian Mesa site in Alaska. A radiocarbon dating by the decay counting method was carried out on scattered charcoal and indicated that the site had a radiocarbon age of 8260 years BP or about 9130 calibrated years BP. Although this considerably post-dates the Alaskan Mesa and Nenana sites it was the first time such fluted points had been reported from Siberia. Vance Haynes, describing the fluted point as an interesting find, noted that it could have been made in America and traded across the Bering Strait. He remarked that there was probably more trading going on across the Bering Strait back then than there was between the USA and the Soviet Union during the 'cold war' [7]. The authors of the article are careful, however, not to claim that Uptar had any direct link with palaeoindian traditions in the Americas. Clearly this is but the beginning of a fascinating attempt to establish whether or not America was initially the eastern part of a vast subcontinent, named Beringia, that connected the Old to the New Worlds and was peopled by humans moving from the Old to the New Worlds across the Bering Strait.

Prior to the report by King and Slobodin, the question became further complicated when a paper by Anna C Roosevelt, a professor of anthropology at the University of Chicago, and a number of colleagues was published in the 19 April 1996 issue of *Science* [8, 9]. A total of 56 carbonized plant specimens from a cave in the Brazilian Amazon from a stratum containing numerous stone points and tools were radiocarbon dated, 49 by AMS at the Lawrence Livermore National Laboratory and 7 by conventional dating. The dates obtained suggested that dwellers of the cave called Caverna da Pedro Pintada at Monte Alegre ('Happy Mountain' in Portuguese) in the Brazilian Amazon near Santarem in northeast Brazil may have lived at the same time as the people of the Clovis site in New Mexico based on the oldest AMS date they obtained of about 12 240 calibrated years ago BP (the radiocarbon dates ranged from 11 145 to 10 230 years BP). Haynes and others, however, think this oldest date may be off by some 700 years and that a more reasonable age for the Brazilian cave dwellers is 11 600 calibrated years ago BP (10 500 radiocarbon years BP) based on an average of 14 of the most accurate earliest dates. That would make it younger than both Folsom and Clovis. Whether the palaeoindians at this Brazilian site represented a culture separate from those at Clovis is still not settled. In any case one of the more surprising aspects of the Monte Alegre discovery is that it challenges the assumption held by many archaeologists that tropical forests did not provide enough food to support people.

Haynes, in the Libby retrospective [3], mentioned a site of human habitation in Chile for which there was claimed to be evidence of artefacts with radiocarbon ages of 11 800 to 13 600 years BP. Haynes noted that one of the artefacts (not susceptible to radiocarbon dating) was a well-made bifacial stone projectile point. The site in question was Monte Verde, some 575 miles south of Santiago, Chile and, as the crow flies, some 9500 miles from the Siberian–Alaskan land bridge where, on the Alaskan side, sites of human habitation of similar, but somewhat younger, ages had been authenticated. If true it was a startling discovery. It would mean that the inhabitants of Monte Verde could not have crossed into Alaska and walked an ice-free path through Canada. Haynes, however, was initially sceptical of the claim.

Monte Verde lies at latitude 41°30′S and longitude 73°15′W in the province of Llanquihue at the southern limit of continental Chile approximately 30 miles east of the shores of the Pacific Ocean. Further south, complex waterway channels and islands stretch some 600 miles to Tierra del Fuego. It seems a most unlikely site for one of the earliest habitations of people in the Americas, raising as it did questions of how its inhabitants could possibly have got there.

The incredible saga of the meticulously careful studies of the archaeology of Monte Verde has been published by the Smithsonian Institution Press in two volumes edited by Tom D Dillehay [10, 11]. Summary articles on this stunning archaeological achievement have also been published recently in *Science* [12, 13]. In his preface to volume I [10] A L Bryan, an anthropologist from the University of Alberta, asserts that 'our concept of early man in the New World will be forever changed by the results from Monte Verde' because it shows that 'most currently accepted assumptions and concepts are untenable'.

Nearly 30 radiocarbon ages were obtained from charcoal, wood and ivory material [13], and of these 18 are listed in volume I of the Smithsonian publication [10]. These 18 samples were divided between the Radiocarbon Laboratory at the University of Texas at Austin which received ten, the AMS laboratory at Oxford University which received two and Beta Analytic, Inc. which received six. Oxford measured the radiocarbon in amino acids from the collagen extracted from an ivory gouge and obtained a radiocarbon age of 12 000 ± 250 years BP and in a wood artefact yielding a radiocarbon age of 12 450 ± 150 years BP. These correspond to calibrated ages of 13 100 and 13 550 years BP respectively. The measurements of other cultural material measured by the other two radiocarbon laboratories are of similar ages with similar uncertainties. The oldest cultural object was some charcoal measured by the University of Texas laboratory that yielded a radiocarbon age of 13 565 ± 250 years BP. This corresponds to a calibrated age of about 14 700 years BP. Why the errors quoted are so large is not explained. In general the weights of these 18 samples ranged from a fraction of a gram to several grams. This amount of material cannot be a source of the measurement uncertainties. There is no doubt, however, that the Monte Verde site pre-dates all the oldest sites found so far in North America.

It was mentioned previously in this chapter that it seemed possible that the earliest human visitors to America could have crossed from Siberia to Alaska by boat even if no land bridge existed or, for that matter, could have navigated down the west coast of North America, establishing landing sites on the way. Until recently, however, there was little or no evidence of such voyages down the west coast. The *New York Times* of 6 October 1996 in an article headlined '9,700-year-old bones back theory of a coastal migration' reported the discovery of human bones in one of many low, muddy caves on Prince of Wales Island on the southern Pacific coast of Alaska, some 225 miles southeast of Juneau and 100 miles from Prince Rupert, British Columbia. The bones were discovered by Dr Timothy H Heaton, a professor of earth sciences at the University of South Dakota at Vermillion. They were carbon dated at the AMS facility at the Lawrence Livermore National Laboratory in California using collagen extracted from the bone and converted to graphite. The 9700 year date (actually the number 9730 was quoted by Livermore) is the usual radiocarbon age BP, as described previously, and corresponds to a calibrated age of about 10 830 years ago. There is an additional complication, however. There was definite evidence that the human in question had subsisted on a diet of marine food, such as fish, seals, etc. This means another correction to the radiocarbon age must be applied. This lowers it to 8900 years BP or about 10 000 calibrated (actual) years.

It was pointed out in the *New York Times* article that the journey down the coast by sea was most likely since, even today, it is difficult to make the trip by land. The article also notes that recent geological evidence indicates that the northern end of the corridor between the Laurentide and Cordilleran ice sheets may have opened later than 12 000 years ago and made it impossible for the Clovis people to have come that way. It is true that the 10 000 year age post-dates that of the Mesa site and Clovis man by several thousand years, but the Prince of Wales discovery makes a sea route south for early humans a serious possibility [2].

The radiocarbon datings of bones from the Makarovo-4 site in Siberia, the southern Alaskan site on Prince of Wales Island, the ivory gouge from Monte Verde and others raise the question of the credibility of radiocarbon measurements on bones. After all, is it not possible that bones buried in the ground for thousands of years can be contaminated with extraneous carbonaceous material? In his article in the Libby retrospective Haynes discusses the carbon dating of bones, mostly mammoth and ancient bison bones but, in at least one case, fragments from a human skull collected at various Folsom sites. He expressed some reservations with the results. The problem with bones, as opposed to charcoal, is to be sure that the carbonaceous material in the bone that is used for dating is material that was actually present in the bone when the animal or human died and not from organic or inorganic carbon-containing compounds incorporated into the bone at much later times. Haynes is not alone among archaeologists and anthropologists in his distrust of bone dates. Clearly, however, if one wants to date the earliest sites of humans in America and if human bones are available at the sites that is, ideally, what should be radiocarbon dated. Equally good would be a bison or mammoth bone implanted with a Clovis or Folsom point. Such a find would be rare indeed, but at least one case has been reported.

The vexing question of how to date old bones has received renewed attention with the advent of AMS. Human and animal bone is complex. One of the experts in the radiocarbon dating of bone is R E Taylor, a professor of anthropology at the University of California at Riverside and director of the Radiocarbon Laboratory there. He is an archaeologist/archaeometrist who has studied the intricacies of dating old bones for many years. He notes that carbon-containing compounds in bone tissue exist in two forms, organic and inorganic. The principal indigenous organic component of bone is the protein, collagen, which is made up of a variety of amino acids. The principal indigenous carbon-containing inorganic constituent exists in the form of a carbonate, apatite. In the case of fossil bones the latter can be mixed with calcium carbonate from ground water sources deposited in the bone matrix and the original organic collagen can be complexed with various fulvic and humic organic compounds from the soil in which the bone was buried.

The advent of AMS has made it possible to separately radiocarbon date various organic products contained in fossil bone. Taylor states that there now appears to be consensus among investigators concerning the reliability of bone radiocarbon values. Where appropriate biochemical purification procedures are employed, accurate radiocarbon ages can be obtained on bone collagen in cases where the bones retain a significant amount of collagen. Bones seriously depleted in their original collagen content can yield seriously anomalous ages. In the latter case the use of non-collagen proteins such as osteocalcin appear to be very promising. It is a field that is still under active study. In Taylor's Libby retrospective article [14] he concludes, 'All investigators agree that bone can be a very difficult sample type with which to work— requiring great attention to detail in sample pretreatment and preparation. The capabilities of AMS technology have provided researchers with a very powerful tool, making it possible to develop a detailed carbon-14 "map" of various carbon-containing compounds in a bone sample. An important goal of such studies will be to develop a preparative methodology which would accurately identify organics indigenous to a bone, irrespective of its diagenetic history.'

The importance of such studies was highlighted by the discovery, in California in the 1920s and 1930s, of several almost complete and clearly very ancient human skeletons. If the ages of these skeletons pre-dated that of the Clovis and Folsom sites found much further inland, as some measurements indicated, it could constitute evidence that the first humans in America did not pass through the ice-free corridor that opened toward the end of the last Ice Age from Alaska to points south but actually may have come by sea at a much earlier time.

One of the most interesting of these human remains, the Sunnyvale skeleton, according to Taylor, was excavated in 1973 from an exposed section of the Sunnyvale east drainage channel in the southern portion of San Francisco Bay [15]. An examination showed that the skeleton was a morphologically fully modern, nearly complete female, statistically (in 32 standard measurements and indices) indistinguishable from a local population of females previously dated both by radiocarbon and cultural associations to between 400 and 1600 years ago. A dating technique called amino acid racemization (AAR) gave an age of 70 000 years. According to Taylor, had this age been even approximately correct 'the Sunnyvale hominid would have been one of the oldest directly dated *Homo sapiens sapiens* skeletons, not just in the western hemisphere but in the whole world. It would also have been necessary to explain why a human population had undergone essentially no morphological change over a 70 000 year period' [14].

The term racemization means the conversion of an optically active substance into a racemic or optically inactive form—one in which there is an equal mixture of dextrorotatory (D) and levorotatory (L) substances. Only L-amino acids are usually found in the protein of living organisms. When studied with polarized light these amino acids rotate the plane of polarization of the light to the left. Over long periods of geological time these L-amino acids undergo slow racemization producing D-amino acids. The proportion of D- to L-amino acids in a fossil steadily increases with time. By determining the extent of racemization in a fossil bone its age can be estimated.

In a paper published in *Science* in 1974 [16], Jeffrey L Bada, a scientist at the Scripps Institute of Oceanography in San Diego, California and two colleagues, pointed out that this technique needed much smaller quantities of fossil bone than were required for radiocarbon dating. This, of course, was three years before the invention of AMS when the conventional decay counting method of carbon dating required several grams of bone. In the *Science* paper they reported the ages (as measured by AAR) of several skeletons that were discovered in the mid 1920s at two sites, one near La Jolla and the other near Del Mar, both in California. The measurements suggested that man was present in North America at least 50 000 years ago. It was in 1975 that Bada reported the AAR age of the Sunnyvale 'girl' as 70 000 years [17].

One problem with the AAR method is that the rate of racemization is temperature dependent. There is no accurate record of temperature variations in California or anywhere else on earth that covers the last 70 000 years, spanning as it does most of the last two Ice Ages. It is thus necessary to calibrate the rate constant by AAR measurements in bones in the same or nearby locations whose ages have been determined by radiocarbon dating. For a variety of reasons, not the least of which is the absence of any credible radiocarbon dates on any organic material 70 000 years old, this calibration is problematic. The reaction of professional archaeologists/anthropologists of the Vance Haynes 'school' to the pre-Clovis dates reported by Bada on the various California hominid bones ranged from puzzlement to rejection. What was needed was a radiocarbon date on the amino acids Bada had used for his AAR studies. This became possible after the advent of AMS.

Among the many letters the Rochester AMS group received after publication of our first measurements of carbon-14 in organic materials in 1977 was one from Professor Bada dated 1 August 1978. In it he described his new method of dating fossil materials based on the racemization reaction of amino acids. He enclosed a number of his publications including the one in *Science* alluded to above. He offered to supply us with 15 to 50 milligrams of amino acids isolated from bones. The amino acid fractions were the same ones he had dated by racemization. At that time we were still being very cautious about accepting samples for dating since it was early in the development of the AMS technique and we did not want to make any mistakes that might discredit the method—particularly on important and controversial material like the ones Bada was offering. In my reply I noted that our background was such that we could not date back in time beyond 50 000 to 70 000 years. I suggested that he wait until we had solved this problem or until the AMS facility at the University of Arizona became operational where, I assumed, the problem would not exist. In retrospect I wish now that we had not been so conservative. It turned out to be a very important test of AMS that would have been of great significance to the archaeology/anthropology community. In the event it did not require an AMS system with a background level any better than ours.

A decade later Bada and the scientists at the University of Oxford's AMS facility made a determination of the radiocarbon ages of the same amino acid extracts used in his original racemization studies, in particular those from the Sunnyvale skeleton. These were the same ones he had offered to us. He had providentially preserved them either frozen or as dried residue. The Oxford results [18] indicated that, rather than being 70 000 years old, the skeleton was a mere 6500 years old. About the same time Taylor, with the assistance of scientists at the University of Arizona's AMS facility, dated three organic fractions he had obtained of post-cranial bone of the Sunnyvale skeleton. These direct radiocarbon AMS results obtained at Arizona gave a range of ages between 3600 and 5100 years [15, 19, 20]. Although the AMS results at Oxford and Arizona gave a rather wide spread in the age of the Sunnyvale woman, it settled the question. Her racemization age was too old by a factor of at least ten. All the other Californian skeletons that amino acid racemization dated to before the last ice age were similarly young and considerably post-Clovis. It was a result we at Rochester could have provided ten years earlier. No doubt members of the Haynes faction of anthropologists quietly rejoiced that the first arrival of man in America still post-dated the last Ice Age.

*A map of South America showing the location of the Monte Verde site in Chile. This site could be 14 000 years old, pre-dating all other sites in the Americas. Its settlers could have travelled by sea along the west coasts of North and South America from Siberia or east by sea from some Pacific Ocean location.*

One of the lessons to be learned from this scientific controversy is that real professionals in science are not immutably wedded to the results of their scientific investigations, especially when they are controversial. Although I am sure Bada had considerable confidence in his racemization results he did not let that deter him from putting them to the test when a new and powerful technique like AMS came along. As soon as he learned of the advent of AMS for radiocarbon dating small samples, he gave us at Rochester the opportunity to date his samples. Caution on our part caused us to decline his offer. A decade later Oxford and Arizona accepted the challenge and, quite properly, gained the kudos.

The date of the initial peopling of the Americas is still far from settled, and new candidates for the earliest settlement are announced from time to time. It now seems certain that the Monte Verde site in Chile [10–13] pre-dates the Clovis sites [3] and all the other North American sites as, perhaps, does also the Monte Alegre site in Brazil [9]. So far there are no reliable dates of humans in the Americas older than about 13 500 radiocarbon years BP (about 14 600 calibrated years ago BP). Radiocarbon dating by AMS is currently limited to organic material that is no older than about 65 000 years although it should be possible to date back as far as 100 000 years. The limit of 65 000 years was obtained at Rochester in 1977 and still applies today at laboratories like Oxford, Arizona, Toronto and others that are specifically designed for radiocarbon dating by AMS. In any case there remains lots of leeway to credibly date humans in America much further back in time than has been done so far. All that remains is to discover sites that people occupied in the Americas that pre-date Monte Verde. The argument that Asians crossed into Alaska from Siberia some 13 000 calibrated years ago BP and made their way south through the ice-free passage between the Cordilleran and Laurentide ice sheets to the southwestern USA is still a plausible scenario. Voyages south along the coast are also a real possibility, as evidenced by the site on Prince of Wales Island. They were not the first, however. That honour currently rests with the settlers of Monte Verde. How humans reached there and from whence they came is still a mystery. If they came by land from North America why have no sites of ages comparable to Monte Verde been found? If by sea why have no other sites near the west coasts of North and South America been found except for the one on Prince of Wales Island and perhaps the one discussed below? On the other hand they could have sailed across the Pacific from some South Pacific island. These questions remain a major challenge for archaeologists.

Another and quite different challenge faces archaeologists and anthropologists engaged in studies of the early peopling of the Americas, particularly of North America. American Indians have been concerned for some time that graves of their ancestors were being 'desecrated' to further such studies. Indian is the word that Columbus gave to the North American natives and it is what most Indians call themselves [21]. In 1990 the federal government passed legislation entitled the 'Native American graves protection and repatriation act' (NAGPRA). This act states that human remains must be turned over to a culturally affiliated tribe.

In the summer of 1996 what eventually became a 90% to 95% complete skeleton of a man was found on the banks of the Columbia River near Kennewick, Washington some 230 miles from the river's mouth on the Pacific Ocean [22]. The land on which the skeleton was discovered is owned by the Army Corps of Engineers and is leased to the city of Kennewick. The skeleton had two fascinating aspects: a projectile point was embedded in his pelvis and, morphologically, his skull did not look Native

American—his features were caucasoid. Collagen was extracted from a sliver of bone from a finger by Donna Kirner who runs Professor R E Taylor's laboratory at the University of California at Riverside. From this she prepared the graphite target for carbon dating at the AMS laboratory at the Lawrence Livermore National Laboratory. It was some 9000 years old. The University of Arizona archaeologist C Vance Haynes, an expert on palaeoindians, noted that such a find was exceedingly rare and of great importance. Scientists at the University of California at Davis think the finger bone sliver is so well preserved that they may be able to extract and analyse DNA. Such a test could answer the question of to what group of people Kennewick man belonged. Was he Caucasian or an Asian ancestor of American Indians? All the oldest skeletons found in North America have been in the latter category and, in fact, new genetic data suggest that the earliest Americans came from Asia [23]. However, the scientists involved were dismayed to find that the opportunity to study the skeleton further might vanish when it was learned that the Confederated Tribes of the Umatilla Indian Reservation claimed it was an ancestor and proposed to take possession of it for reburial on 25 October 1996 under the auspices of NAGPRA. Their claim was accepted by the Army Corps of Engineers, which has jurisdiction over the skeleton. The notion that anyone would claim that a 10 000-year-old human skeleton was an 'ancestor' is remarkable but, at least in this instance, is regrettably true. A thoughtful analysis of the question of the repatriation of archaeologically valuable human skeletal remains has been published recently [24]. The authors conclude that, in the absence of a clear association with the living, there should be a bias against handing human skeletal material over to indigenous groups.

In early July 1997 the US District Court in Portland, Oregon ruled against the Army Corps of Engineers handing the bones over to the American Indian tribes for reburial [25]. Wisely the court opened the door for future research, recognizing that scientists have a legitimate right to have their proposed studies considered. Meanwhile the bones remain locked in a vault under the control of the corps. A suit has been filed by a group of scientists for permission to further study the bones [26]. Adding elements of farce to the situation, the Umatilla tribe insisted that cedar boughs be placed in the box with the bones, following some arcane tribal burial practice, and a group of Norse pagans, claiming the Kennewick Man as their ancestor, were allowed to view the bones.

**Table 7.1.** *The initial peopling of the Americas.*

| Site<br>Country (state) | Radiocarbon<br>years BP | Calibrated<br>years BP | Reference |
|---|---|---|---|
| Clovis<br>(MT, WY, SD,<br>CO, OK, NM, AZ) | 11 200–10 900 | 12 300–12 000 | [3, page 364] |
| Folsom<br>(MT, WY, CO, NM,<br>most accurate NM) | 10 900–10 200 | 12 000–11 300 | [3, page 364] |
|  | 10 890±50 | 11 900 | [3, page 358] |
| Nenana,<br>Alaska | 11 800–11 000 | 12 900–12 100 | [3, page 368] |
| Mesa,<br>Alaska | 11 600–9730 | 12 760–10 830 | [4] |
| Monte Verde,<br>Chile | 13 000–12 500 | 14 100–13 600 | [10, page 142] |
| Monte Alegre,<br>Brazil | 11 145–10 230 | 12 240–11 330 | [9] |
| Prince of Wales Island,<br>Alaska | 9730–8900<br>(marine) | 10 830–10 000 | *New York Times*,<br>6 October 1996 |
| Uptar,<br>Siberia | 8260 | 9130 | [6] |
| Kennewick,<br>Washington |  | 9300 | [23, 24, 25] |

If the court permits, the DNA tests will be carried out by Frederika Kaestle, a graduate student in the laboratory of David G Smith. Smith is an anthropologist at the University of California at Davis. A favourable court decision will also allow all the data on Kennewick Man to be published [27].

Table 7.1 lists the early sites in North and South America, their ages in radiocarbon years BP and their calibrated ages in years BP, i.e. calendar years before 1950.

# Chapter 8

# The American Indians, the Vikings and Columbus

We have seen in the previous chapter that the oldest site of human habitation so far discovered in the Americas is located, *mirabile dictu*, at the southern limit of continental Chile at Monte Verde almost 600 miles south of Santiago. It dates to between 13 600 and 14 100 calendar years before present. This date is some 1000 years older than the oldest sites in Alaska and the southwestern United States. How the people of that ancient Chilean site got there is a mystery. If they came from Siberia they made an epic 9500 mile journey. It must have been mainly by sea, at least along the west coast of Canada, because, from Alaska to the present Canada–USA border from British Columbia to Saskatchewan and further east, it seems likely that there was one solid unbroken sheet of ice until some 14 000 years ago [1]. Some time between 14 000 and 12 000 years ago an ice-free corridor had opened up between the Laurentian and Cordilleran ice sheets [1] that would have permitted journeys by land from Alaska south through Canada. It should be noted, however, that University of Arizona archaeologist C Vance Haynes, a leading expert on palaeoindians, is not convinced beyond a reasonable doubt of the age of the Monte Verde site [2].

We turn now to the more recent history of humans in America following the end of the last Ice Age some 7000 years ago. American Indians occupied much of North America. It was they who developed agriculture and constructed major earthen mound complexes for purposes unknown.

My only direct involvement in carbon dating an ancient Indian site came as a result of my interaction with an undergraduate student in the Departments of both Anthropology and Chemistry at the University of Rochester, Nicholas J Conard. Nick was employed part time at the University of Rochester's Nuclear Structure Research Laboratory, of which I was the director. He was a member of our AMS group. Although he was involved in all aspects of our AMS research which, at that time, involved measurements of a number of cosmogenic radioisotopes including chlorine-36, iodine-129 and carbon-14 in a variety of samples, it was the carbon dating that interested him the most.

Shortly after we made the first measurements of carbon-14 in natural organic matter in 1977 and received much publicity for doing so we began to receive requests from many sources to radiocarbon date a variety of artefacts. By the end of 1979 these numbered four dozen or so, including the one from Bada described above.

Nick came to my office one day in 1982. He explained that, because of his interest in specializing in anthropology in his further university studies, he would like to get involved in some anthropological carbon dating project that exploited the power of AMS. I gave him the list of requests we had received and invited him to choose one that seemed particularly interesting to him. The one he chose had come from Dr David L Asch of the Archeobotanical Laboratory, Center for American Archaeology, in Kampsville, Illinois in August 1978.

Asch was a member of a team of archaeologists investigating important Neolithic Indian sites at Koster, Napolean Hollow, Crane and other nearby sites in west central Illinois. (See the colour section.) He had a 40 milligram sample of carbonized maize found at a level in this dig carbon dated from charcoal obtained at this level to be about 8800 years old. Since this pre-dated by at least a millennium the previous earliest evidence for maize anywhere in America he was particularly interested in having the maize itself dated. The sample size was much too small for conventional decay counting dating—only AMS could do the job. Both Nick and I thought it would be a significant and interesting project.

The material Nick received from Asch included, in addition to the maize sample from this oldest level and several from younger ones in nearby sites, samples of squash rinds and of oil-bearing seeds. We measured two of the squash rinds that came from Koster and Napolean Hollow with our Rochester AMS facility. We obtained radiocarbon dates of $7100 \pm 300$ and $7000 \pm 250$ years

*The people involved in AMS measurements at Rochester of material from a Neolithic Indian site at Koster, Illinois. They are from left to right: M Rubin, H E Gove, N Conard and D Elmore.*

BP respectively. These correspond to calibrated ages of 7950 and 7835 years respectively. Asch and his archaeological colleagues, who provided the samples, had concluded that the squash rinds came from a cultivated variety of cucurbita pepo squash and were evidence for prehistoric horticulture in Illinois.

The maize sample that Asch was so interested in and which came from a stratum that was probably about 8000 years old gave a disappointing AMS radiocarbon age of 500 years at most. Clearly it had worked its way down from much younger strata or had been carried down by some burrowing creature. It reminded me of the apocryphal story of the American tourist who was visiting some ancient Egyptian tomb. She spotted a small object on the floor which she surreptitiously snatched up and secreted in her purse. Back in the States she sent it to a noted Egyptologist telling him of its source and asking him whether he could identify it. After some time he wrote back saying, 'Madam: This object is a result of some creature that crept into the crypt, crapped and crept out.' Obviously not all objects found in ancient sites are ancient.

This was an excellent example of the power of AMS. Because it requires sample sizes of milligrams it has great specificity. One can obtain dates on exceedingly small objects of interest rather than the bulk material that surrounds them. We published the results of our AMS measurements in a March 1984 issue of the

*The Iva annua plant whose seeds contain an oily, edible kernel. The size of the kernels suggested domestic cultivation some 8000 years ago.*

journal *Nature* [3]. It was entitled 'Accelerator radiocarbon dating of evidence for prehistoric horticulture in Illinois'. Conard was the senior author.

In 1966 seeds, stalks and rinds of cucurbita squash and seeds of the *Iva annua* plant, showing evidence of domestication, were found in a Mesoamerican cave in Oaxaca, Mexico. The samples were too small to be radiocarbon dated by the only method then available, decay counting by the Libby technique. The samples were carefully saved, however, and recently some of them were reanalysed by B D Smith, director of the archaeobiology programme at the National Museum of Natural History, Smithsonian Institution, Washington, DC. He arranged for them to be dated by AMS. Nine samples of these seeds and other cucurbita squash material gave ages ranging from 9000 to 7000 radiocarbon years BP (10 000 to 8000 calibrated calendar years) [4, 5]. These are between 50 and 2000 years older than the squash samples found at the site in Illinois [3]. The latter reference was not mentioned in Smith's publication [5]. This caused two of the senior co-authors of the Conard *et al* paper [3] to query Smith.

In reply Smith noted that, 'In the 13 years since the *Nature* article [3] was published, there has been considerable research done on the initial development of domesticated plants in the eastern woodlands of the United States. One of the most interesting discoveries, which has been accepted by almost everyone...is that cucurbita pepo was independently domesticated in eastern North America from a wild indigenous gourd, not from an introduced domestic plant. All of this has been reported in the literature, and debated at great length, over the past decade, so it was neither necessary, nor appropriate, to reference the *Nature* article, since while at the time of publication it appeared to provide early evidence of domesticated squash in eastern North America, that assertion has been subsequently refuted.'

The earliest construction projects of American Indians (Native Americans) were earthen mound complexes. Of these the oldest so far discovered is located at Watson Brake near Monroe, Louisiana [6, 7]. It was reported in the 19 September 1997 edition of the *New York Times* by their senior science writer John Noble Wilford in an article entitled 'Study of ancient Indian site puts early American life in new light'. The site comprises 11 mounds and connecting ridges covering an area of some 900 feet in diameter or 22 acres (90 000 square metres) in area. The heights of the mounds range from 15 to 25 feet. AMS radiocarbon dates on charcoal and humates (organic acids from decayed vegetation in the soil) from levels just below the base of the mounds suggested that mound construction at Watson Brake began between 5400 and 5300 calendar years BP. This is nearly 2000 years older than any previously discovered mound constructions. Previously the earliest mounds discovered were located at Poverty Point, Louisiana, and they were 3500 years old. Thousands of more recent mounds have been found scattered through the eastern United States. Their dates range from 100 BC to AD 1500.

The purpose of the mounds is completely unknown. No human burials have been recovered nor do they seem to have any religious or ceremonial significance. The time and labour devoted to the construction of such monumental architecture was very substantial, but to what end? The study [6] of the Watson Brake mounds was led by J W Saunders, a professor of geosciences at Northeast Louisiana University in Monroe. He was quoted [7] as saying: 'I know it sounds awfully Zen-like but maybe the answer is that building them was the purpose'.

So much for America's settlers of Asian origin—who were its first European visitors? Although it is certain in many people's minds that the Vikings discovered North America hundreds of years before Columbus, there had never been incontrovertible proof. Probably the most complete account of the ancient Norse in Greenland and North America has been written by the renowned Canadian author Farley Mowat in his book *West Viking* first published in 1965 and then as a paperback in 1973 [8]. Mowat has extensive knowledge of northern Canada, of Newfoundland and of seamanship in Canada's northeastern waters. His book is based on the original Norse saga sources and on his own extensive research. It is his account of how what may be the first Viking settlement site in North America came to be that will be outlined briefly later in this chapter. If the settlement described below is not the first it is the only one that has been reliably carbon dated as a late tenth or an early eleventh century European settlement. The most convincing radiocarbon measurements were made at the University of Toronto's AMS facility in 1986 following preliminary AMS measurements in 1978 at the University of Rochester.

In the late 1950s a Norwegian author and explorer Helge Ingstad [9] was convinced by his readings of the Icelandic sagas that the land the Vikings had discovered and called Vinland was located somewhere on the eastern coast of Canada or the USA. He started his search in Rhode Island and gradually moved north, finally arriving in Newfoundland in 1960. Here, on the northern tip of part of Canada's tenth province, Ingstad was led to a site on Epaves Bay (Bay of Wrecks) containing ruins that the local inhabitants assumed had belonged to early colonial settlers. The site is also near a bay adjoining Epaves called, on old charts, Méduse Bay, named after a species of jellyfish, the medusa, that swim into it in droves during the summer months. The bay is now called L'Anse au Meadows, from the French anse or cove and the English corruption of Méduse as Meadows. It is known locally as Lancy Meadows. Etymology can be convoluted. Ingstad preferred to name the site after L'Anse au Meadows rather than Epaves, the bay it was actually on.

The head of the Radiological Dating Laboratory at the Norwegian Institute of Technology in Trondheim, Reidar Nydal, has been involved in the radiocarbon dating of charcoal at the L'Anse au Meadows site for many years. According to him [10], from 1961 to 1968 Ingstad made seven archaeological visits to the site during which his wife, Ann Stine Ingstad, directed the archaeological work. The excavations disclosed structures that had once been turf-walled buildings. No evidence of roof coverings remained but they were likely also turf on wooden pole rafters. Nydal states, 'According to archaeological assessment, the

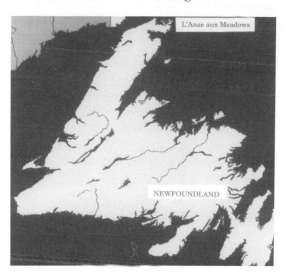

*Map of Newfoundland showing the location of the Norse site at L'Anse au (or aux) Meadows on the northernmost tip of the island.*

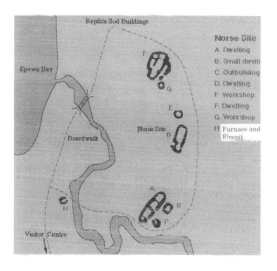

*Layout of the L'Anse au Meadows Viking site at the northern tip of Newfoundland. The location of the iron smelting fire pit is marked as H on this map. The stream flowing into Epaves Bay is Black Duck Brook.*

excavated houses were Norse and similar to types in Iceland and Norway from c. AD 1000. The most convincing proof of Norse origin came from artefacts of the Viking age, such as a stone lamp, a spindle whorl and a bronze pin'. It should be explained that a spindle whorl acts like a flywheel spinning a wooden shaft to twist raw wool into yarn. The one discovered at L'Anse au Meadows was made of soapstone and is taken as evidence that there were women living at the site. Cooking fire pits and a fire pit clearly used for the smelting of iron were also discovered. Nydal, using conventional decay counting radiocarbon measuring techniques at his laboratory, dated the requisite relatively large samples of charcoal and peat found at the site. On the basis of the measurement of many samples he concluded that the date of the L'Anse au Meadows settlement lay in the range AD 975–1020 with the most probable date being AD 1000. Although these dates look pretty convincing, how can one prove the charcoal came from Viking fires rather than those of the indigenous native people? The proof came from a smelting fire pit, and it came from the application of radiocarbon dating by AMS.

Small inclusions of charcoal from the wood fuel used to melt the iron were found in the slag in this pit. This charcoal was carefully picked out bit by bit from the slag by our Toronto collaborators until a milligram or so was collected. A preliminary measurement of its age was carried out at Rochester in 1978 by the General Ionex–Rochester–Toronto consortium and yielded a late tenth century date. In 1986 the slag charcoal material was dated much more carefully by the Isotrace facility at the University of Toronto. Their date was AD 984 to 1010 or AD 997 ± 13 [11]. This extraordinarily small error is due to the steep slope of the dendrochronology calibration curve at this date. The Viking sagas say Lief Eriksson discovered Vinland around the year AD 995. Could L'Anse au Meadows be Vinland? As mentioned below, that is disputed by Farley Mowat. Be that as it may, the fact that it was an iron-smelting fire pit meant it was used by Europeans. Iron, at that time, was an unknown element to the Native Americans, who were still in the stone age, so the iron smelting pits were of European and hence likely of Viking origin.

Nydal describes later measurements at Toronto of samples from the L'Anse au Meadows site in a talk he gave at the Thirteenth International Radiocarbon Conference in Dubrovnik, Yugoslavia in 1988, which was later published in the proceedings of that conference [10]. I attended that meeting and it was a stellar affair. Dubrovnik is a medieval walled city and seaport on the Adriatic, of unsurpassed beauty. Two or three years later it was subjected to a savage bombardment by Serbian forces that began the hateful internecine warfare that has been waged in that troubled part of the world to this day.

At that meeting in Dubrovnik in 1988 I gave a talk [12], to be mentioned in a later chapter, preceding the one by Nydal. It summarized the status of the radiocarbon dating of the Turin Shroud. A little over a month before, I had been present at the AMS facility at the University of Arizona where they had dated the shroud. I could not reveal the results because two other AMS laboratories that had also received shroud samples were still in the throes of making the measurements. It was an awkward position to be in. Every delegate at the meeting was intensely curious about the answer but I managed to avoid revealing what it was.

In his talk Nydal described the L'Anse au Meadows measurements he had made and then described the AMS measurements made by the University of Toronto group led by R P Beukens, one of the Toronto collaborators involved in our early AMS work at Rochester. He said they had measured four samples. These were in addition and subsequent to the measurements they had made on carbon nodules from the iron slag which Nydal, for some strange reason, did not mention. He stated, 'The samples were carefully collected by Parks Canada, who is now in charge of the site, to avoid systematic errors. Their objective was to find wood that had been associated with metal tools, which were not used by the indigenous people'. One sample consisted of the outer portion of a partly charred stick, 1.5 centimetres in diameter, two others consisted of twigs and the fourth came from the outer rings of a tree stump. The first sample was several hundred years younger than expected but the latter three gave a calibrated date of AD 1000–1020 in excellent agreement with the AMS date for the charcoal inclusions in the iron slag and, for that matter, with Nydal's own results obtained by decay counting. Europeans, probably Vikings, had settled the site 500 years before the celebrated discovery of America by Columbus.

Who were these early Viking visitors? In his book. *West Viking* [8] Farley Mowat notes:

Although it has not been stated as a fact, the leaders of the Ingstad expedition have strongly intimated that the Epaves Bay site was the location of Leif Eriksson's Vinland. This deduction is not consistent with the facts as I understand them and as I have set them out in this book. If we are to accept Epaves Bay as Lief's Vinland we must discard, or seriously distort, the description of Vinland as given in the Greenlanders Story—and this is the only description of Vinland we possess.

It is my belief that Epaves Bay was the site of a temporary Norse settlement established there by the combined Icelandic–Greenlandic expedition led by Thorfin Karlsefni and Thorvald Eriksson.

Mowat asserts that Lief's Vinland lay 'in the lagoon of Tickle Cove Pond at the foot of Tickle Cove Bay'. The latter, in turn, can be found at the foot of Trinity Bay in southeastern Newfoundland. Mowat, with some asperity, writes that 'some doltish bureaucrat has recently decided that these names do not carry sufficient dignity'. Tickle Cove has been renamed Bellevue and the pond has become Broad Lake. It was a change Mowat refused to abet! Lief's arrival at Tickle Cove is given in the sagas as about AD 995. As for L'Anse au Meadows, according to Mowat, 'Thus on a day in the late summer of the first decade of the eleventh century, men, women and children began making their way ashore to establish what was to become the first European settlement in the New World'. This was the expedition led by Karlsefni and the date agrees with the AMS radiocarbon dating of L'Anse au Meadows. Some ten years or so earlier Lief merely over-wintered at Tickle Cove so it could hardly be classed as a settlement.

Before leaving the subject of the Vikings and Vinland it is worth recounting the amusing case of the Vinland Map. For some of the information in this account I am indebted to three colleagues, Professor T A Cahill, director of the Crocker Cyclotron Laboratory of the University of California at Davis, Dr G Harbottle, senior scientist in the Chemistry Department of Brookhaven National Laboratory (BNL), Long Island, NY and Professor D J Donahue, co-director of the NSF Accelerator Arizona Mass Spectrometry Facility at the University of Arizona. An excellent account of many of the fascinating aspects of the Vinland Map's history, written by Jennifer Kaylin, appeared in the May 1996 edition of the magazine *Yale* [13]. Three months before, Yale University Press published a new edition of *The Vinland Map and the Tartar Relation*. On that occasion an article by John Noble Wilford, headlined 'Disputed medieval map held genuine after all', appeared in the 13 February 1996 edition of the *New York Times*.

The story began in 1957 when a New Haven rare book dealer bought the map from an Italian bookseller for $3600. He showed his purchase to two staff members of the Yale

*The western half of the Vinland Map showing an island labelled Greenland and another labelled Vinland. The former is quite an accurate depiction of Greenland. The latter may depict parts of Labrador and Newfoundland.*

Library. They concluded, but not without reservations, that the map had been drawn some 50 years before Columbus's discovery of America by an unknown European cartographer and was based on Viking discoveries made some 500 years before Columbus. The map depicted an island in the northwest Atlantic closely resembling Greenland and another, further to the west, occupying approximately the position of parts of Labrador and Newfoundland and labelled 'Vinlanda Insula' that could correspond to Vinland of the Norse sagas. The map was later sold to Paul Mellon for $1 million dollars, and he donated it to Yale.

The map had been bound together with an unrelated medieval manuscript called the *Tartar Relation*. Quite by chance, Yale librarians, during an examination of a medieval encyclopaedia entitled *Speculum Historialia*, comparing watermarks and worm holes in the latter with those in the map and the *Tartar Relation* realized that all three documents had been part of the same book.

*The same as the previous map with only outlines of the land masses included.*

Eight years of research followed that convinced scholars at Yale and others that the map really was a genuine pre-Columbian document and, amidst considerable publicity, the book *The Vinland Map and the Tartar Relation* was published in 1965.

Quite coincidentally, early in 1965 President Lyndon B Johnson had proclaimed 9 October as Lief Eriksson Day (it allegedly corresponded to that noted Viking's birthday). The director of the Yale University Press, Chester Kerr, in an unwitting blunder of monumental proportions, announced the publication of the book containing the Vinland Map on 11 October 1965. He said he had planned to do so on 9 October but, that being a Saturday, he thought Monday was a more felicitous choice. Monday 11 October 1965 was Columbus Day! The Italian-American community was outraged. The ensuing and entirely unmerited notoriety made the book a best seller.

Lingering doubts about the map's authenticity, raised at a 1966 conference on the map held at the Smithsonian Institution, prompted Yale to commission Walter J McCrone of McCrone Associates in Chicago to examine the map. McCrone's expertise lies in the field of microscopy and he has a remarkable penchant for both iconoclasm and publicity. My contacts with Walter were in connection with the Turin Shroud, the putative burial cloth of Jesus Christ. When a series of tests were carried out on the shroud in the fall of 1978 McCrone determined that that there were traces of iron oxide powder on the shroud image. He immediately

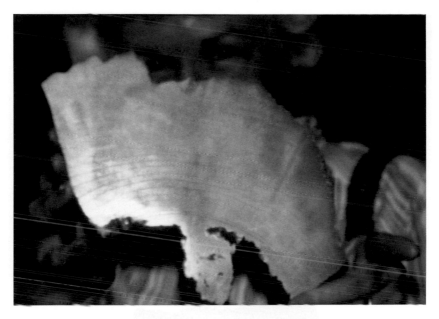

*The author's wife Betty Gove holding a section of a bristle cone pine showing the annual tree rings. The wood is from the Tree Ring Laboratory at the University of Arizona.*

| $^{36}Cl$ | ANTARCTIC METEORITES in collaboration with J.R. Arnold, K. Nishiizumi, M.T. Murrell, and R.C. Finkel (UCSD) |
|---|---|

Chlorine-36 was measured in ten Allan Hills and two Yamato meteorites with sample sizes ranging from 12 to 224 mg. Activities derived from the $^{36}Cl/Cl$ ratios and carrier weights ranged from 5 to 24 dpm/kg meteorite. Comparison with the expected activity and previous results for $^{10}Be$, $^{26}Al$, and $^{53}Mn$ yielded terrestrial ages ranging up to 0.7 million years and a two stage history for one meteorite that we expect broke up in space about 1.3 million years ago.

*An example of how meteorites are found lying on the ice surface in Antarctica.*

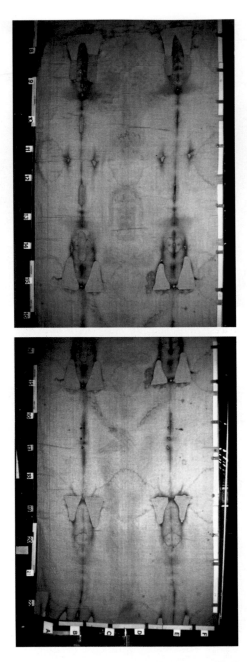

*Photographs of the shroud taken in 1978. Top: note the image of a man's head, front and back, in the centre of the photograph. Bottom: note the crossed hands showing only four fingers on each and the bloody nail hole in the wrist can be seen. The significance of these features is discussed in [1], Chapter 1.*

*Egyptian ibis mummy belonging to Dr L A Garza-Valdes. Dates to between 30 and 330 BC. Courtesy of L A Garza-Valdes.*

*Linen wrapping and bone samples were removed from the ibis mummy by A R David, Keeper of Egyptology, the Manchester Museum, University of Manchester, England. She gave them to D J Donahue, co-director of the AMS laboratory at the University of Arizona, where they were carbon dated by AMS.*

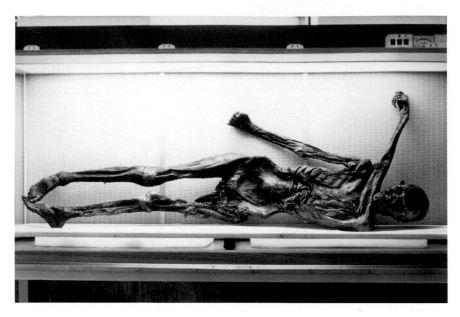

*A photograph of the Iceman stored in a freezer at the University of Innsbruck. Courtesy of Konrad Spindler.*

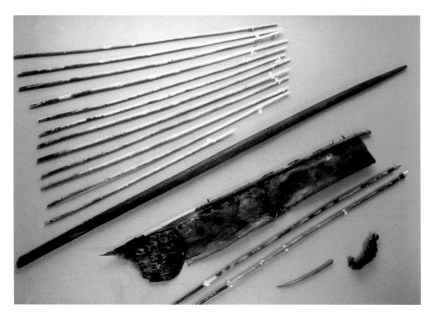

*A photograph of the Iceman's unfinished bow, quiver, arrows, and an irregularly wound cord (probably a bowstring). Photograph: Christin Beeck, Römisch-Germanisches Zentralmuseums, Mainz.*

*A photograph of the Iceman's copper axe. Photograph: Christin Beeck, Römisch-Germanisches Zentralmuseums, Mainz.*

*Photograph of one of the drill sites (the one labelled 12.3 in the figure on page 189). The rusting steam engine was hauled in from the coast over 1000 kilometres away in the early 1900s to drill a well almost a mile deep. It was abandoned when the drilling was completed.*

*A water sampling operation in progress in the Great Artesian Basin.*

*An aerial photograph of the Meteor Crater in Arizona. It is 1.1 kilometres across and 200 metres deep.*

announced that he 'had some good news and some bad news. The bad news is that the shroud is a fake. The good news is that no one is going to believe me'. It remained, however, for others to settle the question of the shroud's authenticity, as recounted in my book *Relic, Icon or Hoax? Carbon Dating the Turin Shroud*, and to do so with somewhat greater objectivity and with a great deal more credibility.

McCrone found that some of the white ink on the Vinland Map contained the pigment anatase. Anatase (titanium oxide) was not used as a white pigment before 1917. The news that McCrone believed the map to be a fake was greeted with relief and joy in Italy and in the Italian-American community, but officials at Yale were not amused. They did, however, concede that the map might, indeed, not be genuine.

In 1985 doubts were raised concerning McCrone's conclusion that the Vinland Map is a relatively recent fake. A friend of mine of many years, Thomas A Cahill, director of the Crocker Cyclotron Laboratory of the University of California at Davis, employed the proton beam from that cyclotron to bombard a portion of the map. In traversing the map the protons generate x-rays that have energies that are characteristic of the chemical elements present in that part of the map through which the protons pass. In particular, in passing through the white ink x-rays characteristic of titanium are generated whose intensities, however, indicated that only trace amounts of that element were present—amounts quite comparable with those found in other medieval documents Cahill had tested. In his opinion, the finding of titanium oxide in the map's paint does not prove that it was drawn after 1917. Titanium oxide is a common chemical compound and is widely distributed in nature. At a conference he and I attended in 1989 he told me that the parchment on which the map is drawn was, very probably, of medieval origin dating to the fifteenth century. He said, however, that McCrone continues to vigorously disagree. I asked him whether it was going to be carbon dated by AMS and he said he had suggested that a couple of years ago but it had not happened yet.

On 10 February 1996 a symposium was held at Yale to discuss the map. McCrone had not been invited to attend but he showed up anyway. He remained silent throughout but at the end of the conference he handed out copies of a talk he would have given, if invited, entitled 'The Vinland Map, still a 20th-century fake'. The map made a brief appearance at the symposium covered by a protective layer of plastic and under armed guard. It is insured for $25 million and is normally stored in a library vault.

Looking at reproductions of the map it is clear that a small amount of parchment sufficient for carbon dating by AMS could be removed from the lower edge of the map, well away from any ink lines or printing, and that is what happened either during or shortly after the symposium. Some 30 milligrams of parchment were cut from the lower right-hand corner under the supervision of Gar Harbottle (BNL) and Jackie Olin (Smithsonian) for dating by Doug Donahue at the University of Arizona AMS facility. It turned out that a waxy coating covered the parchment that needed removing before the radiocarbon age of the parchment itself could be determined. This was accomplished by washing with acetone and then completely evaporating the acetone away. The parchment date so obtained lay between AD 1420 and 1490. Indeed Columbus could have used the map to guide him in his AD 1492 voyage! Whether this settles the question of the authenticity of the Vinland Map is not obvious. Forgers can find blank sheets of parchment of almost any age and there is not nearly enough carbon left to date the ink, even by AMS [14]. In any case that is where the mystery of the Vinland Map stands at present.

The following is another example of the application of AMS in dating ancient American Indian artefacts. Starting in 1911, a cave near Lovelock, Nevada was mined for commercially valuable guano (bat dung). The miners of the material, used as fertilizer, had been alerted to the cave by local Paiute Indians and after most of the guano had been removed Indian relics began to appear. In 1924 a remarkable discovery was made. A package wrapped in rush matting and covered with ashes and reed refuse was discovered in a pit lined with stone in a corner of the cave. It contained eleven duck decoys made of rushes. The decoys, which resemble canvasback drakes, were about a foot long. (See the colour section.) Their bodies were formed from a couple of dozen large bullrush stems bound in a tight hairpin and trimmed at the ends to resemble a duck's tail. Sewed fast to the body was a billed head formed from split reeds. Duck decoys are highly prized collectable items and some examples can sell for tens of thousands of dollars. These ones are on display in New York City's Museum of the American Indian. Until 1983 the museum was reluctant to establish their age. A substantial part of them would have to have been destroyed to carbon date them by the conventional Libby decay counting method. With the advent of

AMS in 1977 it was now possible to do so using only a snippet of reed. The measurement was made at the AMS facility at the University of Arizona. They turned out to be 2000 years old and represent the oldest duck decoys in the world [15].

Undoubtedly there will be many more archaeological discoveries related to the first peopling of the Americas, to the contributions made by the American Indians and to the first Europeans to settle in America. Carbon dating by AMS will, without doubt, contribute to the authenticity of these discoveries. Until then one can say, as things stand now, the earliest people to enter North America crossed from Siberia to Alaska 12 000 to 13 000 years ago. Sites in the southwestern states in the USA and in Brazil are some 12 300 years old. One site in Chile is likely to be at least 13 600 years old, possibly indicating that its inhabitants travelled there by sea, perhaps across the Pacific. It is now widely agreed that the first Europeans to settle in North America were the Vikings around AD 1000. The discovery of America by Columbus occurred 500 years later. American Indians began the practice of agriculture some 10 000 years ago and created large-scale earth structures over 5000 years ago. Our certainty in these facts rests in no small part on the astonishing power of radiocarbon dating by AMS.

# Chapter 9

---

# Nuclear Power, Nuclear Weapons and Nuclear Waste

In 1899 Ernest Rutherford, later Lord Rutherford, a native of New Zealand, became a professor of physics at McGill University in Montreal, Canada. His chair and a physics laboratory at McGill had been endowed by the head of the Macdonald Tobacco Company of Canada. It was there that Rutherford studied the decay chains of uranium and thorium—studies that earned him the Nobel Prize for chemistry in 1908. Macdonald, despite the wealth he had earned from the sale of tobacco products, really did not approve of smoking. As a major benefactor of physics at McGill he was in the habit of visiting the Macdonald Laboratory from time to time. Rutherford smoked a pipe and so the scientists working in the laboratory kept a weather eye out for Macdonald's visits and made sure they aired the laboratory prior to his entry. One day a series of open containers of thorium metal lined one wall of the room and sensitive radiation counters lined the other. Rutherford was smoking his pipe near the thorium pots when a lookout spotted Macdonald walking toward the laboratory. Quickly the windows above the pots were opened as well as the windows above the counters at the other end of the room. By chance the movement of air was from the thorium to the counters and it carried the tobacco smoke across the room. When it arrived at the counters they began to click away merrily. Rutherford instantly surmised that the thorium was emitting a radioactive gas. This thorium emanation was later named thoron. It is now known to be an isotope of the gas radon. Serendipity often plays an important role in scientific discoveries. It is unlikely that Macdonald earned any encomiums for the discovery.

In 1911, while working at the University of Manchester in England, where he had moved from McGill, Rutherford suggested an exceedingly clever experiment involving the scattering of alpha particles from a radioactive source by a thin gold foil.    His brilliant theoretical interpretation of the data showed that the gold atom consisted of a very small positively charged central core surrounded by a much larger halo of electrons of equal but opposite negative charge.    Undergraduate physics majors at American universities still feel hard put upon when required to understand Rutherford's elegant scattering theory.    It is not, however, beyond their ken.    Eventually what became known as Rutherford scattering was used to prove that the central core or nucleus is 100 000 times smaller than the electron halo but it contains 99.98% of the atom's mass.

This epic discovery constituted the birth of nuclear physics. Its importance greatly surpassed that of Rutherford's studies of the decay series of uranium and thorium and should have earned him a second Nobel Prize, this time for physics, but it never did.    Some eight years later Rutherford carried out the first artificial nuclear transformation and made the suggestion that the nucleus contained protons whose number equalled those of the surrounding electrons as well as neutral particles. The existence of the latter was finally established by a colleague of Rutherford's, James Chadwick, in 1932. Chadwick was awarded the Nobel Prize for physics in 1935 for his important discovery of these particles he called neutrons. They had almost the same mass and size as protons but had no electric charge.

I am still surprised and somewhat chagrined to recall that the chemistry textbook I used in my last year of high school in 1940 made no mention of the neutron in its description of the atom. It seemed that news of scientific discoveries took a remarkably long time to reach the secondary school panjandrums of Ontario, Canada. That may have been one of the reasons, when I finally decided on a scientific career a year or so later, why I chose physics rather than chemistry as my major concentration of study—a decision I have never regretted. At Queen's University in Kingston, Ontario, which I attended from 1941 to 1944, I actually learned about neutrons at first hand from one of my physics professors, J A Gray. He had what must have been one of the very earliest neutron sources that he let us use to make artificial radioactivity.

Rutherford recognized, from the beginning of his studies of the atomic nucleus, that it was a potential source of an enormous amount of energy if ever a way could be found to tap it. He doubted a way would be found. However, a clue as to how to tap it came with the discovery of nuclear fission by Otto Hahn and Fritz Strassmann in Germany in 1938. Hahn's collaborator over a period of 30 years, Lise Meitner, then in Copenhagen, along with her nephew, Otto Frisch, came up with an interpretation of the fission phenomenon and discussed it with Niels Bohr, head of the Institute of Theoretical Physics in Copenhagen. Bohr, who had worked with Rutherford in Manchester at the time of the latter's discovery of the nucleus, had developed the theory of the energy levels of the electron orbits around the atom in 1913 after his return to Copenhagen. Bohr's theory of the atom earned him the Nobel Prize for physics in 1922. He was quick to realize the importance of Hahn and Strassmann's discovery of fission and carried news of it to the United States in 1939, just before the start of World War II. There its military potential was immediately recognized by most physicists in America and Europe. Many of the latter had fled the fascist terror in Germany and Italy.

During 1961–62 I was on leave from Atomic Energy of Canada at Bohr's Institute for Theoretical Physics in Copenhagen. My wife Betty and I had the great privilege of meeting and being entertained by Bohr and his charming wife Margrethe at the mansion they occupied on the grounds of the Carlsberg Brewery. This residence, by tradition, was donated by Carlsberg for the use of whoever happened to be Denmark's most renowned citizen. At that time Niels Bohr was unarguably that person. It was popularly rumoured among the Danes that a pipeline connected the mansion directly to the brewery—if it did not it should have! At one of the Bohrs' dinners, held for the scientific visitors to the Institute, my wife and I were fortunate to be seated with Niels Bohr at one of the many small tables in the winter garden of the mansion. Although Bohr's command of English was excellent, he spoke very softly and his accent was virtually impenetrable. Danish friends told me that, even when speaking Danish, he was exceedingly difficult to understand. Knowing the role he had played in bringing the information on fission to the United States and his subsequent involvement with the Manhattan Project (the giant USA wartime enterprise that developed the atomic bomb) I asked him whether he would ever write his memoirs, including his association with this project. His reply was that he would only do so after the most controversial people involved in the Manhattan Project were dead.

Unfortunately Bohr himself died in 1962 shortly after our return to Canada.

On his 1939 trip to America Bohr and his physics colleague and fellow passenger on the Swedish-American liner *Drottningholm*, Leon Rosenfeld, arrived in New York City at 1 p.m. on 16 January. While Bohr remained in New York to conduct some business, John Wheeler, a theoretical physicist and professor at Princeton took Rosenfeld to Princeton. That very evening Rosenfeld described the discovery of fission by Hahn and Strassmann to members of Princeton's physics department journal club. He created a sensation and news of the discovery immediately began to circulate widely in the USA as it already had in Britain. However, many questions remained to be answered before it could be established whether nuclear fission could be harnessed to produce electrical energy or, more importantly, with war looming, to produce a nuclear bomb.

Natural uranium contains two isotopes, uranium-238 and uranium-235, the latter being some 138 times less abundant than the former. Which of the two was responsible for fission when it captured the slow neutrons employed by Hahn and Strassmann? It turned out to be uranium-235. How many neutrons were emitted during the fission process? If the number did not exceed one no chain reaction, necessary for both the production of power and for a nuclear explosion, would be possible. If the number was two or greater (it turned out for uranium-235 to be about 2.5 on average for slow neutron fission) what was the best way to moderate them or slow them down so they, in turn, could initiate further slow neutron induced fissions? Ultra pure graphite or, even better, heavy water was found to be the answer. Perhaps the most important question, one that lay uppermost in the minds of the nuclear scientists in Britain and the USA—many of whom were refugees from Nazi Germany—was how much did their opposite numbers in Germany know about fission and its potential for producing the ultimate horror weapon, the atom bomb? As it developed they did not know nearly enough—and those who might have known and who were not enamoured of the Nazis clearly dragged their feet.

On 1 September 1939 Germany invaded Poland and two days later Britain and France declared war on Germany. The race to beat Germany to the atom bomb began in earnest. On 7 December 1941 the Japanese bombed Pearl Harbor and the next day President Roosevelt declared that the United States of America was at war not only with Japan but Italy and Germany as well. One year later, on 2 December 1942, under the supervision of the Italian-born nuclear physicist, Enrico Fermi, the world's first nuclear reactor went critical. The controlled release of energy from the atomic nucleus was a reality. This first reactor, called by Fermi CP-1 (Chicago Pile number 1), had been constructed in a squash court of the University of Chicago's Stagg Field. It was built of almost 800 000 pounds of pure graphite, 81 000 pounds of uranium oxide and 12 400 pounds of uranium metal arranged in a complex configuration designed by Fermi.

The uranium was of normal isotopic concentration with 138 atoms of uranium-238 to each atom of uranium-235. Only the latter contributed to the energy-releasing fission process that resulted when the uranium-235 captured a slow neutron. Although 2.5 neutrons are emitted each time uranium-235 captures a slow neutron and undergoes fission not all of them contribute to further fission by their capture by other uranium-235 nuclei. Many are lost in capture by impurities in the graphite or simply escape from the reactor assembly. Others are captured by uranium-238 changing it into transuranic elements (elements of higher atomic number and weight). One of the most important of these is plutonium-239 which, like uranium-235, undergoes fission when it captures both fast and slow neutrons. It turned out that for every slow neutron captured by uranium-235 in CP-1 to produce fission, 1.0006 neutrons were available to produce further fission events after they were slowed down by collisions with the carbon atoms in the graphite moderator. That was enough. In the few minutes Fermi permitted the reactor to operate the neutron intensity doubled every 2 minutes. A cadmium rod with a capacious appetite for slow neutrons was then inserted in the reactor and shut it down. The era of controlled nuclear power with all its promises and problems had begun.

Enrico Fermi was a modest and apolitical man but he was a giant in the field of nuclear physics. I met him only once and it was at a party held by some close nuclear physics friends of mine at the time of the annual spring meeting of the American Physical Society in Washington, DC. It must have been in the mid 1950s when I was still working at Atomic Energy of Canada in Chalk River. This APS spring meeting was, and still is, devoted chiefly to nuclear physics, and practically everyone in North America working in the field attended it. At that time the USA was going through another of its periodic phases of communist paranoia.

The Board of Regents of the University of California had initiated loyalty oaths that had to be signed by all faculty members of that great university system on pain of dismissal. These oaths proclaimed that, not only were those willing to sign loyal to the USA, but that they had never been a member of the Communist Party. I and most of my friends in Canada considered the oath quite ridiculous. I found Fermi sitting on a chair, surrounded by a coterie of acolytes seated on the floor at his feet. They were all young nuclear physicists like me, many of whom I knew well, and they were discussing nuclear physics with the great man. During a break in the conversation I had the temerity to ask Fermi his opinion of the California loyalty oaths. It seemed to me that a person as important and influential as Fermi should be willing to take a position on this important political issue. Fermi seemed a bit taken aback by the question but he refused to respond. He simply rose to his feet and walked away. The looks I got from his ardent admirers were withering indeed.

After Fermi's reactor went critical in December of 1942 the next three years were devoted to the development of the atom bomb. This was carried out in the Manhattan Project under the direction of the great theoretical physicist J Robert Oppenheimer and Army General Leslie R Groves. It involved most of the physicists in the United States, many from Britain and many from other European countries who had fled to the States to escape Hitler's terror. It was the subsequent use of two of these atomic bombs in August 1945 in the destruction of Hiroshima and Nagasaki that ended the war with Japan and, simultaneously, ended World War II. These first and only uses of nuclear weapons in war and some of their consequences have been described in Chapter 6. Mankind now faced the uncertain prospects of living with nuclear energy, nuclear weapons and the long-term legacy of nuclear waste. Even to those most knowledgeable in the field the future seemed daunting. There is a bright side, however. Since the atom bombs were dropped on Japan almost 50 years ago no nation has employed nuclear weapons in the many wars that have been waged since then. In terms of the generation of electric power nuclear reactors have proven to be, with two exceptions, safe and reliable. Those exceptions are the 1979 Three Mile Island accident in the USA and the Chernobyl (USSR) nuclear power plant accident in 1986. The radiation released in the latter accident killed or injured many in the immediate area and its effects are still being felt.

Unlike power plants employing fossil fuel, mainly oil and coal, nuclear power reactors contribute virtually nothing to the production of acid rain or to global warming. They do produce high-level nuclear waste and major efforts to safely store such waste are under way as will be mentioned later.

It is now some 40 years since the world's first nuclear reactor for the production of electric power was commissioned. Remarkably that was in the USSR. Initially nuclear power was touted as an infinite source of electricity at practically no cost—a promise that was never realized. None the less its development pressed forward. In 1990, according to the International Atomic Energy Agency, the world's annual nuclear power generation accounted for 16.4% of the world's total electricity generation. This is equivalent to over half the electric energy that could be produced annually at the present rate by crude oil from the Middle East. Burning oil for electric power is not only dangerous to the environment but the supply is not limitless and its sources are politically insecure. The oil shortages of the 1970s and the Gulf War of 1992 dramatically attest to that. At the end of 1990 there were 426 nuclear reactors, each with gross outputs of 30 megawatts (electric) or greater, scattered throughout the world. Of these 112 are in the United States [1].

Eventually the uranium-235 in the nuclear fuel elements in these reactors is largely 'burned' up and new fuel rods must be installed. The fission products in the used rods are highly radioactive and the used fuel must be stored in some safe place until this lethal activity dies away. That can take many hundreds or thousands of years. In addition, these spent fuel elements contain plutonium-239. Since it can be used to make nuclear weapons and can also be used to enrich the uranium required for nuclear reactors it is extracted from the spent fuel (mainly so far from reactors operated for the US Department of Defense) along with any remaining uranium-235.

The reprocessing of the latter fuel for weapons is conducted by the US Department of Energy. The reprocessing of commercial power-plant fuel rods was halted indefinitely in 1977. As a consequence the spent fuel rods from commercial reactors are currently stored in cooling pools near the reactors and await permanent disposal. These on-site storage pools are rapidly reaching their maximum capacity. Until quite recently the reprocessing of fuel from US Department of Defense reactors, including those used to power submarines and warships, was carried out at a number of Department of Energy sites throughout the USA including the Hanford Reservation in Washington, the Savannah River site (SRS) in South Carolina, and the Idaho National Engineering Laboratory (INEL) in Idaho. Reprocessing operations at all three sites have now been closed down. West Valley in New York State was, until recently, the only privately owned reprocessing plant.    It was used to extract residual uranium-235 and plutonium-239 from used commercial power reactor fuel rods.  By presidential order it ceased such activities in 1976.

However, these four sites still store about 11 million cubic feet (0.3 million cubic metres) of high-level nuclear waste distributed between them.   The total amount of this waste would fill a football field piled 228 feet deep according to a 1985 publication issued by the American Institute of Professional Geologists [2]. Unknown amounts of this toxic radioactive waste are leaking into the environment at all four sites and are reaching and exceeding plant boundaries through ground water that eventually leads to nearby streams and rivers that flow through inhabited areas. It has all the potential of a nightmare scenario and is certainly presented as such by public foes of nuclear energy for weapons or electric power.  Any technique that can monitor this nuclear waste leakage, in principle, should be of great interest to the US Department of Energy.  The US DOE and the management of these four sites are actively exploring ways to securely contain this enormous volume of waste and ways to incorporate it into materials that resist leaching so that it can be stored safely in some, as yet to be determined, permanent disposal site. The magnitude of the task is daunting but a beginning is being made [3].

Some time in 1987 my colleagues and I in the accelerator mass spectrometry (AMS) group at the University of Rochester's Nuclear Structure Research Laboratory were approached by Dr Thomas M Beasley concerning a project of great interest to him.   Beasley is a geochemist working at the US DOE Environmental Measurements Laboratory in New York, NY. He had for some years been familiar with AMS and its potential for measuring radionuclides at exceedingly low levels in a variety of environmental samples. In fact, prior to joining the US DOE laboratory in New York, he had been employed by the Argonne National Laboratory near Chicago and had tried, without success, to establish a full time AMS laboratory there using a tandem Van de Graaff accelerator which was soon to be decommissioned as a nuclear physics research instrument. Tom is a person of great wit and charm with a low tolerance for sham and pomposity. From time to time his minor diplomatic contretemps and his insistence on stating facts in an unambiguous and straightforward manner cause him some problems in the Byzantine organization that is the US DOE. He is, however, a very intelligent and zealous scientist who has made notable contributions to the solution of environmental and other problems that continue to afflict the US DOE.

For some time he had been considering how one might go about measuring the leakage of waste from nuclear fuel reprocessing plants at SRS, INEL, Hanford and West Valley. It occurred to him that a measurement by AMS of chlorine-36 in ground water at these places might do the trick. He knew that compounds of chlorine are not deliberately incorporated into reactor fuel. However, chlorine is an abundant and ubiquitous element in nature and so there were bound to be stable chloride impurities associated with the reactor fuel elements during the time the fuel was fissioning in the reactor. Stable chlorine-35 has a high probability for capturing thermal neutrons and converting to chlorine-36. The latter radioactive element with its half-life of 301 000 years is the one we had been the first AMS group to measure in environmental samples such as ground water and that we also measured in construction material from Hiroshima as recounted in Chapter 6.

At the fuel reprocessing plants the fuel elements are dissolved in hot nitric acid as a prelude to the extraction of plutonium-239 and whatever remained of uranium-235. The dissolved liquid, in addition to containing long-lived fission products, probably also contains chlorine-36 and any leakage of this witches' brew into the environment would release chlorine-36 into the ground water flowing through the site. Airborne emissions from plant operations would also contain chlorine-36 which could fall out into surface streams. The chlorine-36 itself would pose no health hazard but might provide a tracer for more lethal material that might accompany it.

Beasley asked us whether we would be interested in collaborating with him on such measurements to be made initially at the SRS, provided he could persuade the management of that nuclear fuel reprocessing site that it would provide them with valuable information. It might, of course, turn out to be information that they would rather not have. In any case I felt that this was an excellent example of a role that a university based laboratory could play as a general service to the nation. We agreed to work with him provided any results obtained could be published in the open literature. Beasley would collect the water samples, undertake the careful chemistry to extract the chlorine in the water as silver chloride in a way that reduced its sulphur content (and thus the amount of stable sulphur-36 that would mask the chlorine-36) to a minimum and bring the silver chloride to Rochester. Using our AMS facility we would measure the chlorine-36 to stable chlorine ratios in the samples. Somehow Beasley persuaded the US DOE and the management of SRS to cooperate. Maybe they were responding to the biblical axiom, 'The truth shall set ye free'.

Since 1951 the SRS had been operated by the DuPont Company for the US DOE to reprocess nuclear fuel and also to produce tritium for use in hydrogen bombs. With its inconvenient half-life of 12.3 years the tritium content of the most fearsome of all nuclear weapons must be topped up on this relatively short time scale. The huge operation involved in the SRS brought unheard-of prosperity to the region in terms of both money and culture—as the US DOE was proud to boast whenever criticisms of the safety of the enterprise surfaced. As time went on such criticism surfaced more and more frequently. SRS lies just to the northeast of the Savannah River which it directly borders. Under the Savannah River site lies the Tuscaloosa aquifer, one of the largest underground aquifers in the USA. It supplies drinking water to South Carolina, North Carolina, Georgia and Alabama. Pools of water in the SRS containing radioactive waste lie above this aquifer. Nine tanks of an old design containing high-level liquid nuclear waste from the fuel reprocessing operation are leaking into the ground water, but spokesmen from DuPont and the US DOE claim that the most lethal fission products in this waste, strontium-90 and cesium-137, are 'immobilized'. For 35 years DuPont has buried low-level solid nuclear waste in a 192 acre landfill, much of it in cardboard boxes, as if it were mere household rubbish. Over the same period the company has dumped 200 000 gallons per day of toxic and radioactive liquid waste containing strontium-90 and other fission products into open pits. There are some who say the US DOE and its contractors are now suffering from a high-level hangover from a 40-year binge of building bombs. In April 1989, after 38 years, DuPont withdrew as the operator of SRS without acknowledging an iota of malfeasance. Most of the above information was contained in an hour long television series 'Point of View' that aired on PBS during the summer of 1993. This particular programme was entitled 'Building bombs: the legacy'. It is a documentary that must have caused no little embarrassment to the US DOE. Not enough to make them blush, however.

Beasley proceeded to collect surface water samples both on and off the Savannah River site. From them he extracted the chlorine in the form of silver chloride and brought the samples to Rochester. Altogether we measured some 26 of these surface water samples collected at various distances ranging from 15 to 156 kilometres from the SRS nuclear fuel reprocessing facility. In addition six samples were measured from water collected in streams within the SRS. One of the latter had such a high chlorine-36 to stable chlorine ratio that it was removed as quickly as possible from our ion source to prevent contamination of the system. If nothing else, it provided dramatic confirmation of Beasley's idea that chlorine-36 is released during fuel reprocessing operations. From the complete set of measurements we concluded that the on-site level of chlorine-36 was 10 to 20 times higher than cosmogenic levels and came from SRS operations. Even at distances up to 100 kilometres (60 miles) from the site the chlorine-36 concentrations indicated contributions from SRS operations. The results were published in the journal *Ground Water* in 1992 [4].

In assessing the measurements of chlorine-36 at SRS one had to take into account not only the cosmogenic contributions to this radioisotope in the environment but also another anthropogenic source—the atmospheric nuclear weapons tests that were carried out in areas of the Pacific Ocean in the 1950s. Between 1951 and 1958 some 225 tests of nuclear weapons were made in the Pacific. Of these, 33 of the nuclear bombs were exploded on barges. The neutrons emitted in these explosions interacted with chlorine in the sea water producing chlorine-36. This caused an increase by a factor of almost 1000 in the chlorine-36 concentration in the atmosphere over the cosmogenic level and, in a very short period of time, this pulse of chlorine-36 entered the ground water throughout the world. The pulse itself has a relatively narrow time scale. The chlorine-36 concentration started to rise in 1951, reached its peak of 1000 times the pre-bomb test level between 1955 and 1960 and by 1975 had declined back to the pre-test level. At Rochester our AMS group was the first to map this bomb pulse in a Greenland ice core [5] and in ground water collected at the Canadian Forces Base at Camp Borden, Ontario [6]. Some time

*An index map of Greenland showing the location of the Dye 3 site where the ice core discussed in [5] was drilled.*

later a much more detailed measurement of the chlorine-36 content of a Greenland ice core was carried out by the AMS group in Zurich, Switzerland [7]. The presence of this chlorine-36 bomb pulse in ground water provides a useful way of measuring the flow rate of such water.

If one knows the input source of water in an underground water system one can measure the flow rate by determining the distance between this recharge area and the location of the bomb pulse. When calculations are made for the downward and upward flux of water controlled by infiltration and evapotranspiration the net downward flow rate can be quite accurately evaluated. This turns out to be important for the determination of the integrity of potential radioactive waste disposal facilities. We made such measurements in 1989 [8] in collaboration with scientists from the University of Texas in Austin who were evaluating a site in the Chihuauhuan Desert in Texas as a potential low-level radioactive waste disposal area. Our measurements showed that the net flow downwards of water to a depth of 0.5 metres was only

*A photograph of the Dye 3 installation. Originally an extension of the DEW (early warning radar) line it is now a radio/radar check point. It has complete shop facilities. The structure shown is eight stories high and is constructed on posts to let the snow blow through.*

*An ice driller with a 4 inch diameter core segment being carried to a science work station.*

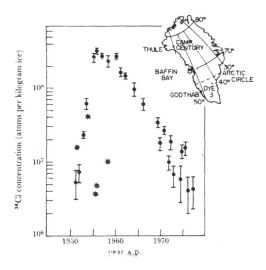

*A plot of the chlorine-36 concentration (atoms per kilogram of ice) in a 100 metre ice core drilled in 1980 at Dye 3. Each sample contained 1 year of precipitation covering the years from 1950 to 1978. The age scale was determined from the $^{18}O/O$ ratio. Also plotted as stars are the major nuclear tests carried out in the Pacific between 1952 and 1958. The chlorine-36 concentration climbed rapidly by about three orders of magnitude above pre-bomb levels from 1950 to 1955, held steady until 1960 and then declined exponentially back to pre-bomb levels with about a 2 year half-life [5].*

1.4 millimetres per year and, thus, the site was suitable for storage of low-level radioactive material. Federal law requires that each state establish a site for the storage of low-level waste. The citizens of this area of Texas are quite happy to accept the suggested location for such waste storage because of the economic benefits. The same does not apply to certain other states, particularly New York State, where the opposition to any site proposed thus far has been vehement.

Returning to the measurements of chlorine-36 at US DOE fuel reprocessing facilities, Beasley proposed that we next investigate the situation at the Idaho National Engineering Laboratory. It is located in southeastern Idaho about 40 miles west of Idaho Falls and is among the largest of the US DOE's nuclear facilities. Originally designated as the National Reactor Testing Station the site was commissioned in 1949 as a complex where prototype nuclear reactors were constructed and tested for military, industrial and research purposes. Over its operating history, 52 reactors have been constructed and, of these, 13 are currently capable of operation. Perhaps one of the most fascinating

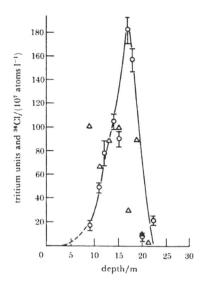

*Chlorine-36 atoms per litre of water (open circles) and tritium concentrations (open triangles) as a function of depth at a landfill site at the Borden Canadian Forces Base in Ontario, Canada. Again the bomb pulse is unmistakable and occurs at a depth of about 20 metres. Although there is not a clear correlation between depth and age the shape is similar to that found in the Greenland ice core where the maximum occurred in 1955. Clearly the chlorine-36 bomb pulse is well suited to studies of hydraulic flow and dispersive mixing. Bomb-produced tritium is disappearing and is now replaced in such studies with bomb-produced chlorine-36.*

test reactors was designed for use in powering an aircraft of a size sufficiently enormous to contain the reactor. The largest runway in the world still exists on the site to this day. It is said that when President Kennedy visited the site in the early 1960s and was shown the prototype of this multistoried engineering nightmare he burst out laughing. Needless to say it never got off the ground.

The INEL is on the eastern portion of the Snake River Plain aquifer. At the northern site boundary the aquifer lies some 60 metres below the surface, while at the southern site boundary it deepens to more than 275 metres. Beasley originally intended to sample freshly fallen snow down wind of the reprocessing facility during times when the facility was operating to demonstrate that atmospheric emissions of chlorine-36 occurred. However, in early 1990, the fuel reprocessing facility at INEL was deactivated for a facility upgrade and since then this activity has been suspended altogether. It was decided to measure chlorine in the deep aquifer itself to see whether any liquid wastes from the plant had leaked or were leaking into the aquifer. Measurements

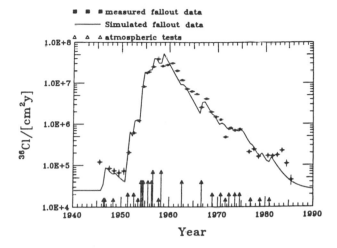

*More detailed data on the fallout of chlorine-36 on the ice sheet at Dye 3 measured by the AMS group in Zurich [7]. Courtesy of H-A Synal, Paul Scherrer Institut, Zurich.*

## Chihuahuan Desert, Texas

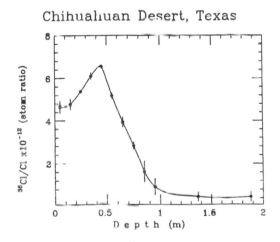

*The depth profile of the $^{36}$Cl/Cl ratio in a bore hole from the Chihuahuan Desert, Texas. The slow downward movement of water (0.5 metres in 35 years) indicates the site would be suitable for low-level nuclear waste storage.*

of chlorine-36 in aquifer water well away from the INEL site showed evidence of only cosmogenic contributions—the bomb pulse chlorine-36 has not yet penetrated through to the aquifer. Measurements of chlorine-36 in on-site aquifer water gave clear evidence that operational activities at the INEL had indeed injected a considerable amount of this radioisotope into the aquifer and it

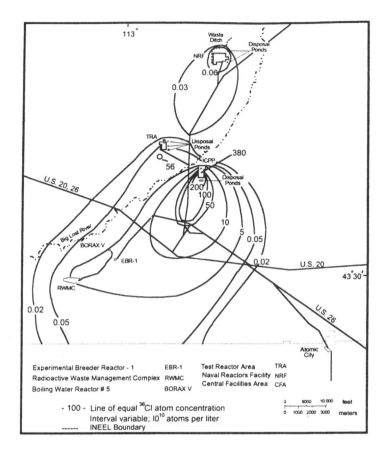

*Isopleths of chlorine-36 concentrations in water in the eastern portion of the Snake River Plain where the Idaho National Engineering and Environmental Laboratory is located (see [9]).*

was reaching and exceeding plant boundaries. That might imply that more dangerous fission products were doing so also. Once again our data were published in the journal *Ground Water* [9].

In mid April of 1993 Tom Beasley and I visited the Idaho Falls headquarters of EG&G, the company that currently has the contract from the US DOE to operate INEL. The visit had been arranged through a good friend of mine, Ralph DeVries, a professor of physics at the University of Wyoming, who was serving as scientific advisor to the General Manager, EG&G, Idaho, Inc. on their INEL operations. Ralph had been a Research Associate in my University of Rochester nuclear physics laboratory during the time we were having such tiresome interactions with

Berkeley over who first invented the technique of accelerator mass spectrometry. He had later served as Assistant Director for General Science, Office of Science and Technology Policy in President Ronald Reagan's White House. Our purpose was to discuss with EG&G officials the possibility of measuring technetium-99 in the Snake River Plain aquifer both within and outside the INEL boundaries.

As mentioned above, chlorine-36 is not a fission product. It is merely a possible indicator of the presence of such products. Technetium-99, on the other hand, is the most abundant of the light fission products produced in the fission of uranium-235 in the fuel elements of power reactors. Technetium is not produced cosmogenically and no isotopes of technetium occur in nature so there is no other source of technetium-99 than from used reactor fuel, at least not in the Snake River Plain aquifer. It should be detectable by AMS and, if so, should make it possible to trace leakage from nuclear fuel reprocessing operations to much greater distances than can be done by measuring chlorine-36 or iodine-129 (which is an abundant heavy fission product). The limits of detection of the latter two isotopes are set by their cosmogenic levels. The EG&G officials expressed interest in the idea and it is currently being pursued.

I found my visit to Idaho Falls fascinating. Although I did not actually visit the INEL, Tom Beasley took me on a tour of the plant boundaries. Tom is a native son of Idaho and knows all the vagaries of the state and its inhabitants. His descriptions of, and anecdotes about, Idaho and its people were very entertaining. It is, of course, largely rural and mostly Mormon. In Idaho Falls a large Mormon tabernacle surmounted by a golden statue of the Angel Moroni dominates the scene. The countryside comprises seemingly endless fields of potatoes, alfalfa, sage brush and cattle. Herds of prong-horn antelope roam through the sage bush and nesting boxes for raptors atop high poles dot the landscape, the latter erected through the good offices of the US DOE.

We stopped at a general store in Howe, Idaho (population 23) to get fuel and some cold drinks. The proprietor, a young farm lad, was taciturn to a fault and clearly suspicious of strangers. When Tom asked him whether the nearby Little Lost River was flowing at the moment he merely replied, 'On and off'. I believe those were the only words he spoke. There were two fuel pumps—one for leaded and the other for unleaded—something I had not seen for some years. They are indicative of the importance of farm

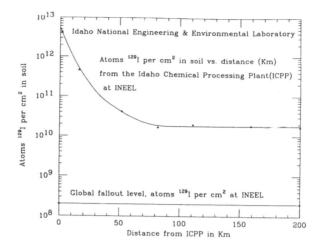

*The distribution of iodine-129 in soil as a function of distance from the Idaho Chemical Processing Plant at the Idaho National Engineering and Environmental Laboratory. The soil samples were collected along a southwest–northeast transect at distances from 1 mile to 124 miles from the chemical plant.*

operations in the area. We paid for the fuel and our drinks and left, much, I am sure, to the young man's relief. Tom said that on a previous visit he pulled his rental Honda into a petrol station only to have the attendant ask him accusingly, 'That's a Jap car, ain't it?' The tone of voice suggested to Tom that an affirmative answer might result in a refusal to sell him fuel, or worse.

He recalled a meeting he had with officials of INEL shortly after we had made some initial chlorine-36 measurements on water from the site. Tom was explaining the significance of the results obtained so far. He said one of the senior managers at the meeting was clearly suffering from narcolepsy. Some of what Tom had been reporting must have seeped through, however, because the official awoke suddenly and said, 'It ain't our chlorine-36, it's the Navy's' (spent fuel from US nuclear submarines is reprocessed at INEL). Tom's courteous but mildly paternalistic reply was, 'Let me lead you through this again'.

In 1994 Tom Beasley and some colleagues at INEL collected soil samples at various distances from 1 to 124 miles from the main radioactive waste storage sites at INEL. The number of atoms of the abundant heavy fission product, iodine-129, per gram of soil were measured at the AMS facilities at the Universities of Toronto and Rochester. The concentrations ranged from a factor

*Seaweed sample collection sites along the west coast of Canada and USA north and south of the mouth of the Columbia River. Courtesy of J C Rucklidge, University of Toronto.*

of 10 000 (at one mile), to 1000 (at 10 miles), to 100 (at 30 to 124 miles) above the background level of iodine-129 created by past above-ground nuclear weapons tests. These measurements substantiated admissions made in December 1993 by the US DOE that millions of pounds of highly radioactive used reactor fuel from reactors employed in weapons production have been sitting in water storage pools at SRS, INEL and Hanford. The cladding of these fuel elements is deteriorating and the fission products are escaping into the water. The concrete basins either unlined or lined with steel containing this water, in turn, are leaking into the environment. The fission product iodine-129 is just one of many that are escaping and one of the least dangerous at that. A comparison of this distribution of iodine-129 as a function of distance from the radioactive waste storage sites at INEL with the distribution from the reactor accident at Chernobyl will be made below.

Another potential site for measurements of chlorine-36, iodine-129 and technetium-99 is the Hanford reservation. It is a federally owned area straddling the Columbia River in south-

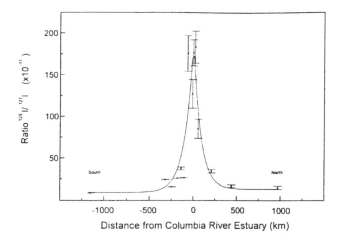

*The ratio of $^{129}I/^{127}I$ as a function of distance from the mouth of the Columbia River. Courtesy of J C Rucklidge, University of Toronto.*

central Washington. Its principal function is to reprocess used fuel elements from plutonium-production defence reactors. It is estimated that the Hanford facility has approximately 800 million litres (200 million gallons) of high-level nuclear waste in storage. Some of it is leaking into the ground water at the site and thence into the Columbia River. In addition there is at least one decommissioned commercial nuclear reactor buried at Hanford. The AMS facility at Rochester has made some preliminary measurements of chlorine-36 samples extracted from ground water from the site but not in enough detail to be of value at present. The Columbia River flows some 300 kilometres (200 miles) past Hanford to its mouth in the Pacific. In April 1992 an enterprising group from the University of Toronto's AMS laboratory collected samples of red and brown seaweed from intercoastal regions along the Pacific coast from Carmel in mid California to Port Angeles in northern Washington state. They had previously collected similar seaweed samples from the lower portion of British Columbia in Canada. These seaweed varieties are known to have high concentrations of iodine. The ratio of iodine-129 to stable iodine was measured by AMS at Toronto [10]. The ratio peaked at the mouth of the Columbia and fell off by a factor of two 75 kilometres (46 miles) on each side. The peak value was almost 20 times that at the extremes at Carmel and British Columbia some 1000 kilometres south and north. Nobody had to guess the source of this iodine-129, an abundant heavy ion product of nuclear fission. The levels of contamination are undoubtedly below those considered as health hazards but the fact that such contaminants are there at all in the Columbia River, an important water resource for the region, is troublesome indeed. Again it took AMS to make the measurements possible and there will be many more to come.

Still another potential site for such measurements by AMS is West Valley, NY, located 35 miles south of Buffalo and operated for the US DOE by a wholly owned subsidiary of the Westinghouse Electric Corporation. It is the site of the only US commercial nuclear fuel reprocessing plant. It operated from 1966 to 1972 and reprocessed 640 metric tons of spent nuclear fuel to recover usable uranium and plutonium. During the operation of the facility, about 600 000 gallons of liquid, high-level radioactive waste were generated and stored at the site. When reprocessing was halted in 1972 there were still unprocessed spent nuclear fuel elements remaining. These are stored in fuel pools on site and measurements of tritium, cesium-137 and strontium-90 in the water of the fuel pools and for at least 100 feet away in water downstream suggest that fission products are leaking from the fuel rods into the pool waters and thence into waters outside the pools. In this connection it should be explained that 'downstream' is with respect to the flow of surficial water through the site. Wells upstream of the fuel pools are not contaminated, but those downstream certainly are. The surface water flows northeast but is thought to be pretty well confined to the site. A potentially more important pathway for the flow of radionuclides from on-site sources would be via Buttermilk Creek that flows through the site northwest 3 miles to Cattaraugus Creek, which, in turn, flows west to Lake Erie.

I visited the West Valley site in July 1993 on my way up to my cottage north of Algonquin Park in Ontario, Canada. The reprocessing facility, fuel pools, intermediate and low-level solid waste sites and surface liquid waste storage pools occupy 200 acres of the 3345 acre Western New York Nuclear Service Center. Driving east to the site along the Schwartz Rock Springs road off Highway 219 that runs south from Buffalo I passed an old farm house with a prominent sign out front reading, 'Welcome to Waste Valley'. In 1992 West Valley employed 738 people. Its principal activity was to incorporate all the high-level liquid waste into non-leachable solid material through a process of vitrification.

Ultimately this glass-like material and the unprocessed fuel rods will be shipped to some as yet to be determined permanent storage site—possibly Yucca Mountain in Nevada.  At that time it is proposed to close the site down.

During my visit I discussed with one of the senior project scientists the possibility of our AMS laboratory at the University of Rochester making chlorine-36 measurements on water from the West Valley site. He was interested but hardly enthusiastic. In his defence I should add that he had many other responsibilities that were fully occupying his time. It was he who told me there were still as yet unprocessed spent nuclear fuel elements stored on the site in fuel pools.  He said there was evidence that radioactive material in the fuel was passing into the fuel pool water and from there into downstream wells.  He carefully avoided using the verb 'leak'.  He said that, at West Valley, 'leak' was a four letter word!  Several months later, after my visit, I still had no indication of interest by West Valley officials in our making chlorine-36 measurements. Clearly they were going to need some convincing that 'the truth shall make ye free'.

The measurements of iodine-129 in seaweed made by the AMS group at the University of Toronto across the mouth of the Columbia River that showed Hanford was leaking radioactive material into the environment suggested that similar measurements should be made as a function of distance from West Valley.  Such measurements can be carried out without requiring any cooperation from officials at that facility. A project to do so was mounted in early 1996 by University of Rochester scientists [11]. They found levels of iodine-129 elevated by more than a factor of 10 more than 125 miles from West Valley. Those levels increase steadily toward West Valley and are 10 000 times above background in two creeks that drain the land surrounding the facility, Buttermilk and Cattaraugus. No wonder 'leak' is a four letter word to the folk at West Valley. The scientists found no evidence that the two nuclear power plants in western New York, Nine Mile Point in Oswego and Ginna near Rochester, were contributing to the elevated levels of iodine-129.

Since West Valley ceased accepting used fuel elements from commercial nuclear power reactors it has been necessary to store their used fuel in water-filled pools at the reactor sites, but such storage space is now filled close to capacity. The spent-fuel storage problem at these sites is critical and a solution must be found soon. The Nuclear Waste Policy Act of 1982 authorized the US DOE to construct permanent geological repositories for spent power plant fuel and other forms of high-level waste. Two such potential storage sites are Yucca Mountain in Nevada and a salt bed site in New Mexico called WIPP (the Waste Isolation Pilot Plant).

The requirements for such geological repositories are stringent indeed. It must isolate high-level radioactive waste from humans for thousands of years, outlasting institutions, governments and nations. The most likely mode of transportation of radioactivity out of any waste burial site is transport by ground water. Thus one must be able to reliably predict water movement around and through the site for 10 000 to 50 000 years. If ground water does contact the waste how much will reach the environment and how soon? The debate over whether the above two sites fulfil these requirements is fierce. At Yucca Mountain the detection of chlorine-36 by AMS in ground water is providing some answers. At salt bed sites such as WIPP the abundant presence of stable chlorine completely masks the presence of chlorine-36. In any case Yucca Mountain is, at present, the site most favoured by the US DOE for the storage of high-level nuclear waste from commercial power reactors.

Yucca Mountain is a region of volcanic tuff—the detritus of volcanic eruptions that occurred aeons ago. It is a barren flat-topped ridge located in the Nevada Desert 100 miles northwest of Las Vegas. It is some 1500 feet above the surrounding countryside and stretches for six miles. The water table lies 1900 feet below the land surface. That would allow the construction of a repository within the mountain entirely above ground water. This contrasts with other candidate sites where the water table is nearer the surface and the repository would be beneath the water table. Furthermore the region around Yucca is extremely arid— the Nevada Desert is one of the driest places in the United States. The average precipitation at Yucca Mountain is 1.5 centimetres a year.

The earliest measurements of chlorine-36 carried out by AMS at Rochester located the bomb pulse at a depth of about 0.5 metres, indicative of a very slow downward movement of precipitation in the tuff [12]. In later Rochester AMS measurements at another site on the edge of the proposed Yucca Mountain repository the bomb pulse was located at a depth of 153 metres [13]. This was taken as an indication that the flow of precipitation was occurring through faults and fractures. If that does indeed turn out to be the case it would cast doubts on the integrity of the Yucca Mountain site as a high-level nuclear waste repository. It is still far too early, however, to jump to this conclusion.

Meanwhile political rather than scientific arguments are causing delays. In December 1995, the US Secretary of Energy, Hazel O'Leary, informed a Senate committee that, although large caverns have been carved out in Yucca Mountain for the storage of spent fuel, its permanent or even temporary storage there is delayed until at least the year 2015 because of spending cuts in the Federal budget and opposition by local inhabitants. Meanwhile spent fuel from nuclear power stations is increasing at an alarming rate with no solution to its permanent storage in sight.

Opponents of nuclear power base their opposition primarily on their fear of reactor accidents, for example the China Syndrome, and secondarily on the dangers of the leakage of high-level nuclear waste contained in used fuel elements even if it can be 'safely' stored. Proponents scoff at such fears and, until 26 April 1986, pointed out that no nuclear power reactor had suffered a major accident. Furthermore, they claimed, the safe storage of nuclear waste was entirely feasible and its implementation was being selfishly blocked by the NIMBY (not in my back yard) phenomenon.

The complacency of nuclear power proponents was severely shaken, if not shattered, when at 1:23 a.m. on 26 April 1986, unit No 4 of the Nuclear Power Station in Chernobyl, Pripyat, Ukraine (about 50 miles north of Kiev) suffered two explosions. The first, when the power level of the reactor rose to 120 times its capacity, blew the containment lid off the reactor and the second, when the reactor fuel and graphite inside the core exploded, sent chunks of graphite and high-level radioactive material from the fuel through the ruptured concrete roof of the reactor building. Vast areas around Chernobyl were instantly contaminated and the winds carried a plume of radioactive particles throughout the world. Although the Nordic countries were most affected, all of Europe suffered and the radioactive fallout even reached the United States. Major evacuations and other emergency procedures were carried out in the regions surrounding Chernobyl and a concrete sarcophagus was constructed to blanket what remained of the reactor. What caused this tragedy?

Unit No 4 was scheduled to be shut down for maintenance. While preparing for the shut-down the operators were instructed to observe the dynamics of the reactor under limited power flow. The disaster occurred as a result of operator error during this unusual mode of operation, despite the fact that the reactor operators were highly trained and regarded themselves as an elite crew. It was concluded that they had become over-confident and did not think carefully about the effects of some of their operating procedures on the functioning of the reactor in abnormal circumstances.

In August 1992 a number of AMS and other laboratories, including Rochester, received a letter from the International Atomic Energy Agency (IAEA) in Vienna, Austria asking whether we would be interested in measuring iodine-129 in soil originating from the Chernobyl region. We were told that the samples came from about 120 miles northeast of Chernobyl and that initial studies showed that the concentration of iodine-129 was $10^9$ atoms per gram of soil (a factor of about 100 above background). As mentioned previously iodine-129 is an abundant fission product, as is its much more lethal but shorter lived partner iodine-131. It is the latter with its higher energy beta and gamma rays, its short half-life of 8 days and its propensity to accumulate in the thyroids, particularly of children, that leads to death by cancer. Iodine-129 merely tells how much iodine-131 was present in the early days following the release of nuclear fuel from an operating reactor. Both isotopes of iodine are present in equal amounts as fission products in the fuel and both are distributed equally when the fuel cladding of an operating reactor is breached in an explosion. The results of the measurements showed that the initial IAEA estimate was correct. Interestingly enough (and quite coincidentally) the iodine-129 concentration at this distance of 120 miles was very similar to that found at a similar distance from INEL in Idaho. More recently the iodine-129 concentration in soil has been measured about half a mile from the disabled Chernobyl reactor. It is around $10^{11}$ atoms per gram of soil and is comparable to that measured at one mile from the radioactive waste storage sites at INEL. At INEL, as opposed to Chernobyl, however, the iodine-129 was never accompanied by iodine-131. The latter had decayed away to negligible levels by the time the waste containment facilities at INEL had begun to leak.

A report in the *Rochester Democrat and Chronicle* of 27 April 1995—nine years after the Chernobyl accident—contained some sobering news from Moscow. The Ukranian Health Minister disclosed that the estimated death toll from the reactor disaster was 125000, with a disproportionate share of deaths among children, pregnant women and rescue workers. The report stated that, 'More than 140 000 people have had to abandon homes rendered uninhabitable by radiation and half a million more continue to live amid the contamination'. It is expected that oncological diseases in people exposed to the radiation will peak in the second decade following the accident (mid 1996 to mid 1997). Just for comparison, the death toll in Hiroshima by the end of 1945 from the bomb that was dropped in early August of that year was about 140 000.

The most serious nuclear power accident in the USA occurred at Three Mile Island's No 2 reactor located near Middletown, Pennsylvania on 28 March 1979. A series of human and mechanical failures caused the cooling system of this reactor with its capability of producing 790 megawatts of electric power to malfunction. This damaged the reactor core and caused a leak of radioactive gases. There was no loss of life or property damage other than that to the reactor. The clean-up took 14 years and cost about $1 billion dollars. Although the accident was serious there was no explosion and no major environmental contamination. Its twin reactor, unit No 1, continues to supply electric power and is scheduled to do so until it is decommissioned in 2014.

Turning to the production of nuclear weapons; it is now recognized by the United States Department of Energy (DOE) and other US authorities that a daunting environmental remedial programme is needed to rectify the conditions created by the manufacture of such weapons in the United States between 1945 and 1986. It is estimated that, during those years, almost 60 000 nuclear warheads were manufactured at various locations throughout the country. A similar number were produced in the former Soviet Union. The volume of radioactive waste resulting from this production of nuclear weapons, principally stored at Hanford, Idaho Falls and Savannah River, with lesser amounts at other locations, is mind boggling. According to the 30 August 1996 issue of *Science* and based on the US DOE's own estimate, 'cleaning up its 3700 sites in 34 states could take $230 billion and 75 years'. There are some who think both estimates are much too low. In that same *Science* report it was noted that the US DOE has taken what it hopes is a major step toward reaching solutions to the massive nuclear waste clean-up problem. Some $112 million over a three-year period will be awarded as grants to university scientists to fund research in new ways to solve the problem. The old attempts at vitrification (incorporating the waste in glass) and pumping and treating contaminated ground water, among others, are rapidly falling into disrepute. What new techniques will replace them in the future is not yet clear, but a start is being made.

We have seen in this chapter the role that accelerator mass spectrometry (AMS) has played in obtaining some answers to questions related to nuclear power, nuclear weapons and nuclear waste.   One thing is clear.   AMS can be used to determine whether the leakage of nuclear waste into the environment is diminishing as a result of whatever clean-up strategies will be employed in the future. Whether it will be so employed is, at present, problematical. Many of us in the AMS field find this apparent reluctance to use such a powerful analytical tool difficult to comprehend.

# Chapter 10

# Carbon Dating the Turin Shroud

Without any question the most fascinating artefact with which I have been involved since the invention of accelerator mass spectrometry (AMS) in 1977 is the Turin Shroud—the reputed burial cloth of Jesus Christ. Soon after the genesis of carbon dating by Willard Libby in the 1940s the question of whether the age of this remarkable cloth could be determined by Libby's method was raised. It is said that Libby himself volunteered to date the shroud but his offer was declined when it was learned that the sample size he would have needed was that of a man's handkerchief. No one was likely to permit the sacrifice of so much cloth from such a precious object. AMS, on the other hand requires sample sizes a thousand times smaller than the method invented by Libby. For the first time establishing the age of the shroud seemed feasible and in 1988 the mission was accomplished. In a previous book entitled *Relic, Icon or Hoax? Carbon Dating the Turin Shroud* (see reference [1], Chapter 1) I recounted the complexities of that mission and I summarize them here. Some of what follows is taken directly from that book.

Until February 1993 the shroud was stored in an elaborate silver casket on the altar of the Royal Chapel of the Cathedral of John the Baptist behind a triply locked iron grille. At that time it was moved to a safer place inside the cathedral itself while the dome of the chapel underwent renovations for forthcoming public exhibitions. Still sealed in the silver casket it was placed in a bulletproof glass case in the cathedral. On 11 April 1997 a disastrous fire broke out in the chapel. Before the fire was extinguished it did extensive damage to the chapel and water and

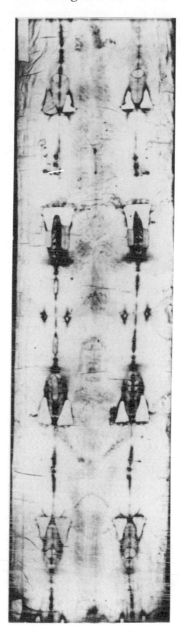

*Black and white photograph of the Turin Shroud. It is a stained linen cloth of herringbone weave the colour of old ivory, 14 feet 3 inches long and 3 feet 7 inches wide bearing the faint back and front head-to-head image of a crucified man. A very noticeable feature is the 16 triangular patches applied by the Poor Clare nuns in AD 1534 after a fire in AD 1532.*

*Location of the shroud until about 1996 when it was moved to a location in the main Cathedral of St John the Baptist. This location is in the Royal Chapel located at the rear of the cathedral. The shroud lies rolled in an ornate silver chest behind this locked metal grille.*

smoke damage to the cathedral itself. Had the shroud still been located in the chapel it undoubtedly would have been destroyed. As it was it was saved by the firefighters and removed undamaged from the cathedral for temporary storage in the apartment of Giovanni Cardinal Saldarini, Archbishop of Turin. Despite the fire it was planned that the shroud be displayed publicly during the month of April 1998 to celebrate the hundredth anniversary of the first photograph taken of the shroud by Secondo Pia.

Prior to February 1993, while the shroud rested in the Royal Chapel, it was only displayed to the public on special occasions every 40 years or so. However, nearby, there was a full-sized, colour photograph of the shroud. It shows an altogether impressive and beautiful stained linen cloth the colour of old ivory, some 14 feet long and about 4 feet wide. (See the colour section.) It bears the faint front and back imprint of a naked crucified man. The image depicts all the stigmata of the crucifixion described in the bible including a large blood stain from the spear wound in the side and from the nail holes in his wrists and feet. The linen weave is a three to one herringbone twill. A seam or tuck divides the main body of the shroud from a 6 inch side strip of the same weave which runs almost the entire length of the cloth. A backing cloth of basket weave covering the entire back area of the shroud is exposed at both ends of this side strip where pieces of the side strip have been either removed or never existed.

What strikes the eye first, upon viewing the shroud, are the 16 patches that line the image. These were applied in pairs to the front of the shroud in 1534 two years after it was damaged in another fire. This one occurred in the chapel in Chambery, France, where the shroud was stored, again in a silver chest. Gouts of molten silver burned through the shroud in a symmetric fashion due to the way in which it was folded in the chest. The shroud was doused with water before the fire damage could spread to the image. This near catastrophe, however, did yield some interesting scientific information. Silver melts at a temperature close to 1800 °F (960 °C). Because the shroud was folded inside the chest, there had to be a considerable variation of temperature at various points on the image ranging from something near this high temperature to ones approaching normal room values. Yet there was essentially no change in the appearance of the image from one region to another. Since many art pigments volatilize at temperatures well below the melting point of silver the ones that could have been used, if it is a painting, are limited. The patches were applied by the cloistered order of Poor Clare nuns. At the same time the backing cloth was added to strengthen the linen.

In 1978 I visited Turin to explore the possibility of carbon dating the shroud by the newly developed accelerator mass spectrometry method requiring very small samples and had the opportunity of seeing it first hand. It was on display in the cathedral and by the time I arrived it had been viewed by close to 3 million people. The shroud, fully unfolded, was mounted on a wall at the end of the cathedral behind bullet-proof glass. At the time Turin was the headquarters of the notorious Red Brigade and every important building was guarded by armed and patently nervous uniformed young men. I found myself strangely moved as I watched the tear-stained faces of the people filing by the shroud. Could this possibly be the burial cloth of Jesus Christ? Could the faint image be Christ's portrait? As a scientist I thought it exceedingly unlikely but I must say I rather hoped that it was. And another thought occurred to me—why should science intrude itself into the life of this lovely object? It was clearly providing religious inspiration to millions of people. Was this not a case where, although science could provide an answer to its age, it need not?

The shroud's known history dates to about the year 1353 when a French knight Geoffroi I de Charny announced its existence. He died in the Battle of Poitiers in 1356 without revealing how it had come into his possession. De Charny was a literate man and the author of the only book on chivalry up to that time written by a layman. He placed the shroud in a church in Lirey, France. It was exhibited around the time of his death amidst considerable controversy. The local bishop felt it was being sold to the faithful as the authentic burial cloth of Christ. He declared that it was an icon and claimed to know the name of the artist—a name, if indeed he did know it, he never revealed. It was exhibited again in 1389 although this time no claim was made that it was Christ's shroud.

The shroud eventually passed into the hands of Margaret, daughter of Geoffroi II de Charny. In 1453, in a complicated fashion, it passed from Margaret to the House of Savoy. In 1502 it was moved to a Savoyard chapel in Chambery, France. In 1578 the seat of the House of Savoy moved to Turin, Italy and the shroud moved with it. It was used on various ceremonial occasions associated with the Savoy family. The shroud remained a possession of this royal family up to the time of the death in 1983 of the last king of Italy, Umberto II, who willed it to the Vatican.

The shroud remained in Turin except during World War II when it was moved to the Abbey of Monte Vergine in the mountains of southern Italy. It since continues to reside in Turin despite Turinese paranoia that the pope will one day order it to be moved to Rome. The Archbishop of Turin has been designated by the pope as the custodian of the shroud and the person who makes all major decisions regarding its care and disposition. At the time the shroud was carbon dated the Archbishop of Turin was His Eminence Anastasio Cardinal Ballestrero. He was succeeded by His Eminence Giovanni Cardinal Saldarini who was Archbishop of Turin at the time of the 1997 fire.

A remarkable event in the life of the shroud occurred in 1898 when an eight day exposition took place. As mentioned above, an Italian photographer Secondo Pia was permitted to take the first ever photographs of the shroud. The equipment of the period required long exposure times under the most brilliant arc lighting during which it is a wonder the image did not pale several shades. Late one night Pia began developing his precious photographic plates. To his wonderment the image slowly emerging on the plates immersed in the developer solution appeared as a positive not, as he expected, as a negative! To him it was as if he were seeing the face of the Lord!

To this day the fact that the image on the shroud is a negative is regarded by the faithful as evidence that it was not the creation of an artist but must have been the result of some mysterious burst of radiation occurring at the time of Christ's resurrection. Consistent with this view is the fact that the purported bloodstains are positives as they should be if the blood from the many wounds depicted on the image soaked into the fabric. Theories have been advanced for a more natural production of a negative image, however.

In 1973 the shroud was shown for the first time on Italian television. A forensic expert from Zurich, Switzerland, Max Frei, was permitted to remove pollen samples using sticky tape applied to the shroud surface. Frei, although not a pollen expert, claimed that some of the pollen came from plants indigenous to Palestine. However, professional palynologists tend to be sceptical of claims that pollen is a reliable indicator of provenance. At the same time two samples were cut by nuns from the hem of the shroud for examination by Professor Gilbert Raes, director of Ghent University's textile laboratory. One sample about 5 square centimetres in area was taken from the main body of the cloth and the other 3 square centimetres in area was cut from the side strip. His examination of the samples under an electron microscope convinced Raes that there were trace amounts of Egyptian cotton present in the predominant linen of the shroud. This could constitute evidence that the shroud was woven on a loom in the Near East previously used to weave cotton.

He is said to have stored the samples in a stamp box in his desk and to have brought them out occasionally for the edification of his guests. A few years later the shroud's custodian, the then Archbishop of Turin, was reminded of their whereabouts and requested their return. Raes shipped them back by ordinary surface mail. Their safe arrival in Turin, despite the vagaries of the postal service, has been facetiously suggested as an indication that, indeed, the shroud must possess some miraculous property. The scientific endeavours of both Frei and Raes were the precursors of a long series of subsequent equally uninformative scientific excursions on the shroud and the nature of its image. The Raes' samples so casually removed for such superficial and pointless examination were large enough for carbon dating by the newly developed AMS technique. However, their ill-documented history while in Raes' keeping was eventually deemed by the science advisor to the Archbishop of Turin to render them unsuitable.

Except for carbon dating, the most recent scientific foray on the shroud occurred at the conclusion of its exposition in 1978. For a period of five days and nights a group of scientists subjected the shroud to a variety of experiments. Their aim was to establish that it was, indeed, the burial cloth of Jesus Christ. They came mainly from US government military and research establishments and had banded together under the aegis of the non-profit organization called the Shroud of Turin Research Project, Inc. (STURP).

They comprised mainly true believers in the shroud's authenticity. They overwhelmed the Turin ecclesiastical authorities and their scientific advisors with their aerospace technology and overbearing military secrecy and discipline. Like all the scientific investigations that had gone before, their final published results were ambiguous and generally of negligible importance despite the time and money expended. STURP's members were so convinced it was Christ's shroud that I was determined to prevent their involvement in its carbon dating. I feared the most important measurement that could be made on the shroud would be rendered less credible by their participation. Fortunately in this I was successful.

Possibly not since the days of Galileo has such a curious interaction between science and religion taken place. It culminated in the only measurement that could provide definitive information on a fundamental property of the Turin Shroud, namely the time at which the flax from which the shroud's linen came was harvested. Nobody would argue that this measurement had any scientific significance unlike many others that have been made by the new method called accelerator mass spectrometry (AMS) and which have been described in previous chapters. However, the wide public interest in the shroud and consequently in any scientific technique that could unambiguously establish its age made it a legitimate object to be tackled by AMS. Funding for the development of AMS came, in general, from US federal sources and that means from tax revenue. If, occasionally, this money is used for projects of not high scientific import but that capture the interest of the general public, that seems to me to be quite proper. The Turin Shroud adventure began shortly after the invention of carbon dating by accelerator mass spectrometry.

The first measurements of carbon-14 in natural samples obtained at Rochester in early 1977 and described in Chapter 2 engendered considerable publicity and, as a result of an article appearing in *Time Magazine* of June 1977, the Rochester consortium received a letter dated 24 June 1977 from the Rev. H David Sox of London, England, then Secretary General of the British Society for the Turin Shroud. He inquired if members of the group had ever heard of the shroud (they had not), and whether it could be carbon dated using the new technique. I replied that the technique was still too new, but that Sox would be kept informed of progress.

About this same time, a group under the direction of Dr G Harbottle at the Brookhaven National Laboratory, Long Island, New York, developed a small proportional counter technique which permitted the method of carbon dating by decay counting to employ similarly small samples. However, because of the low decay counting rate of such small samples, the time taken to make a sufficiently accurate measurement extended to several months.

Somewhat less than two years later, scientists at both Rochester and Brookhaven Laboratories felt their new small sample dating techniques were ready for use in measuring important artefacts. The original offer to date the Turin Shroud using very small samples and using both the tandem accelerator mass spectrometry and the small proportional counter technique was initiated by the group at the University of Rochester and made jointly by them and the Brookhaven National Laboratory. The offer was made by me on behalf of both groups directly to Anastasio Cardinal Ballestrero, Archbishop of Turin in a letter dated 16 February 1979. The offer elicited no response.

As mentioned above, some four months earlier, in October 1978, a team of US scientists, all members of the non-profit organization the Shroud of Turin Research Project, Inc., participated along with some European scientists in a week-long series of non-destructive but largely inconclusive measurements of the shroud. These tests were permitted by Turin ecclesiastical authorities on the advice of Luigi Gonella, the scientific advisor to the Archbishop of Turin, among others, after a prolonged period of complex negotiations between these authorities and the STURP leadership. No member of the STURP team had any expertise in radiocarbon dating. Although I announced the possibility of dating the shroud at Rochester using the accelerator technique at a two-day scientific conference preceding the STURP tests it was made clear by church authorities in Turin that no shroud sample would be made available for carbon dating. STURP, however, has always indicated its strong support for a radiocarbon dating of the shroud. The paper on how we would go about carbon dating the shroud by the new AMS method was published in the proceedings of that two-day conference [1].

After their foray on the shroud STURP formed a carbon-14 dating committee in April 1979 with Robert Dinegar, a chemist at the Los Alamos National Laboratory, as chairman. He approached various carbon-dating laboratories suggested by me concerning their interest in participating in carbon-14 measurements on the shroud. Positive responses were received from six of these laboratories: the University of Arizona, Brookhaven National Laboratory, Harwell, Oxford University, the University of Rochester, and the ETH at Zurich.

The involvement of the British Museum in dating the shroud had its genesis at the 1982 Archaeometry Conference in Bradford, England attended by people representing some of these six laboratories. They suggested that the British Museum be asked to provide them with samples of cloth, with ages known to the museum but unknown to the six groups, for a laboratory intercomparison test. The museum agreed and the results were presented at the Twelfth International Radiocarbon Conference held in Trondheim, Norway, in June 1985. The results suggested that all six laboratories were well qualified to date the shroud [2]. However, as a result of an error in sample preparation by the laboratory in Zurich their measurement of an Egyptian linen sample was about 1000 years in error. This mistake was later corrected. Such an erroneous result (called an outlier) could not have been eliminated on a statistical basis if, for example, only three measurements had been made. Most of the scientists involved therefore felt this underscored the need for more than three laboratories to participate in dating the Turin Shroud.

On 25 June 1985, representatives of the six laboratories which participated in the British Museum intercomparison tests, along with a representative of the British Museum, met informally at the Norwegian Institute of Technology in Trondheim, Norway. The results of the tests had been presented the previous day at the Twelfth International Radiocarbon Conference [2]. I organized this informal meeting to discuss what, if any, the next steps should be that might lead to radiocarbon dating of the Turin Shroud. It was the expressed interest of the six laboratories in dating the shroud that led to the intercomparison tests conducted by the British Museum. A tentative procedure, that I authored, was subsequently agreed upon; it specifically was to involve the Pontifical Academy of Sciences. That academy comprises eminent scientists in many fields and of many religious persuasions. It was formed by some pope many years ago to advise the head of the Roman Catholic Church on matters of science that interacted with religion.

Meanwhile, however, the academy's president Professor Carlos Chagas, had been asked by Vatican authorities independently to assess a proposal by STURP which included as one of some two dozen suggestions for further scientific tests of the shroud, a measurement of its radiocarbon content. When I learned about this I sent Chagas the Trondheim proposal for carbon dating the shroud. It was this proposal that was ultimately considered at a workshop held in Turin on 29 September to 1 October 1986.

Partial funding for the US participation in that workshop was supplied by the National Science Foundation, with myself as the principal investigator of the grant. I assisted Professor Chagas in organizing the workshop, which was authorized by His Holiness Pope John Paul II and sponsored jointly by the Pontifical Academy of Sciences and the Archdiocese of Turin. It was attended by ten representatives of seven carbon dating laboratories who had agreed to conduct carbon-14 measurements on the shroud (one laboratory, the AMS facility at Gif-sur Yvette in France, had been added to the group); a representative of the British Museum, a textile expert from Bern, Switzerland, four representatives of the Pontifical Academy of Sciences and the science advisor to the Archbishop of Turin, Professor Luigi Gonella of the Turin Polytechnic Institute. Five other delegates, three of them from STURP, whose participation was requested by the Archdiocese of Turin, also attended. The meeting was chaired by Professor Chagas, president of the Pontifical Academy of Sciences. The conclusions and procedural steps resulting from this workshop and agreed to by *all* the delegates, including Professor Gonella, were carefully crafted to make the final result of the age of the Turin Shroud cloth as credible as possible. This Turin Workshop Protocol was subsequently published by me in the Proceedings of the Fourth International Symposium on Accelerator Mass Spectrometry held at Niagara-on-the-Lake in Ontario, Canada in April 1987 [3]. That symposium celebrated the tenth anniversary of the invention of AMS and the proceedings appeared in volume B 29 of *Nuclear Instruments and Methods in Physics Research* (see reference [7], Chapter 3).

The first indication that the Turin Workshop Protocol did not meet Gonella's wishes came in an interview with the Turin newspaper *La Stampa* which was published in their edition of 27 April 1987. In particular, he indicated that 'only two or three laboratories' would participate. This and other information I obtained led me to dispatch a telegram to His Eminence Cardinal Ballestrero which was signed by representatives of the seven laboratories and the British Museum.

Despite the warnings in this communication about the likely unfortunate circumstances that would attend the measurements if conducted in the manner proposed by Professor Gonella, the latter persisted in advising the Cardinal to abrogate the original Turin Protocol. On 10 October 1987, the Cardinal sent a letter to all the workshop participants stating that, following the Pope's instructions, the number of laboratories would be reduced from seven to three. He also ordered other fundamental changes in the dating procedure. Among the latter was the decision to prevent the Pontifical Academy of Sciences from playing any further role in the enterprise. There is some reason to surmise that His Holiness' directives were at the Cardinal's request, undoubtedly as recommended by Gonella.

This letter was followed by an invitation from the Cardinal to the three selected laboratories Arizona, Oxford and Zurich. At my urging, a return letter to the Cardinal, signed by all three, carefully explained the possible dangers in having only three laboratories involved, citing the outlier measurement of Zurich in the earlier British Museum interlaboratory comparison measurements [2]. They stated their preference for following the original Turin Workshop Protocol and asked the Cardinal to reconsider his decision. Presumably acting on the advice of Gonella the prelate refused. At a meeting with Gonella in London on 22 January 1988, representatives of the three laboratories and the British Museum quite inexplicably acceded to Gonella's demands and accepted his conditions for a radiocarbon dating of the cloth of the Turin Shroud. One can only speculate about why they abandoned their previously stated view that there was a danger in having only three laboratories involved. Certainly the attraction to date the shroud must have been almost irresistible.

Samples were removed from the shroud in Turin on 21 April 1988 in the presence of representatives of the three AMS laboratories, the British Museum, the Cardinal of Turin and his science advisor Luigi Gonella, a Turin microanalyst Giovanni Riggi (who actually performed the cutting), and two textile experts, one from France and the other from Turin. Although invited to be present as a guest of the Cardinal, Professor Chagas chose not to attend.

The shroud samples and three control samples were encapsulated in encoded stainless steel cylinders by the official from the British Museum and were borne in triumph by the AMS labora tory representatives back to their laboratories. Except for the encapsulation step, all the procedures were videotaped under Riggi's direction. The measurements were carried out at Arizona in May 1988 as described below, and subsequently at Zurich, and finally at Oxford. A copy of the videotape was sold by Riggi to the BBC in the UK. It formed part of an excellent programme produced by the BBC on the carbon dating of the shroud, which included the actual measurements conducted by Zurich without revealing the result they obtained. It would have been much more convenient for the BBC to film the dating practically next door at Oxford but the head of the Oxford laboratory had managed to greatly annoy the BBC people involved by suggesting they pay him for the privilege. Both the Zurich and Oxford measurements were completed considerably later than those at Arizona.

The other person at Rochester who had greatly assisted me in the many complexities of arranging to carbon date the shroud was Shirley Brignall. She serves as administrative assistant to the chair of the Department of Physics and Astronomy and held that position when I was chair from 1977 to 1980. It was during those years that AMS was developed at Rochester and when my involvement with the shroud first began. She and I were the two University of Rochester delegates to the Turin workshop in 1986 and she accompanied me to Tucson, Arizona to view the first dating of the shroud in May 1988. Unfortunately we were informed that she could not be present at that epic event and she was, quite understandably, deeply disappointed. We were told that the Bishop of Arizona had requested permission to attend but even he was politely refused.

I arrived at the University of Arizona's AMS laboratory about 8 a.m. on 6 May 1988. The co-directors of the facility, Professors Paul Damon and Douglas Donahue, had invited me as an observer and I was the only person outside Arizona's AMS group who had been permitted to be present. I remarked that I could not think of another scientific measurement that equalled the one about to take place in terms of general public interest—not, of course, in terms of scientific interest. Perhaps the discovery of the tomb of Tutankhamun was in the same class. There was general agreement

*Those present at the Arizona AMS carbon dating facility on 6 May 1988 when the age of the shroud was determined. They include D J Donahue (third from the left standing), A J T Jull (fourth from the left standing), H E Gove (sixth from the left standing) and P E Damon (seventh from the left standing).*

among those present. It was a remarkable experience for me, who had had a major responsibility in bringing the Turin Shroud to the test of time, to be about to observe the carbon dating take place and to learn before all but a handful of people present how old the shroud really was.

At 9:50 a.m. on 6 May 1988 (Arizona time), the first·of several radiocarbon measurements on a sample from the famed Turin Shroud was completed using the University of Arizona's tandem accelerator mass spectrometry (AMS) facility. It took 10 minutes and consumed only a small amount of an already miniscule sample of cloth Arizona had received. The piece used measured less than 0.25 × 0.25 inches. The measurement proved the flax which formed its linen threads was harvested in AD 1350—the shroud was only 638 years old! Since that year very closely matched the shroud's historical age, it was arguably the least exciting of all possible results.

A month later Shirley and I flew from Toronto to London and on 19 June 1988 we flew to Dubrovnik to attend the Thirteenth International Radiocarbon Conference. The conference headquarters were in a luxurious tourist hotel on the Adriatic Sea. I had been invited to give a paper on the current status of the shroud dating project.

The meeting went extremely well. My talk on 'Progress in radiocarbon dating the Shroud of Turin' [4] was delivered to a capacity audience including Willy Woelfli, head of the Zurich AMS facility, with whom I had several long and pleasant conversations during the meeting unrelated to shroud dating. Donahue and Damon were also present, as well as representatives of the AMS laboratory at Oxford. Zurich were very close to having finished their measurements. In his opening talk in the session I chaired Woelfli was quite critical of some of the experts in the field who suggested that AMS was not really ready to date the shroud and, in any case, there was still not enough experience in AMS laboratories in dating cloth.

Part way along in my talk I stated that, at the invitation of Donahue and Damon, I was present in Arizona when the first measurement was made. I went on to say 'Let me tell you the date...' and when I saw the whole audience lean forward with a collective gasp, I smiled and continued '...the date the first measurement was made'. After the talk, one of my colleagues said he was pleased to see that I was not embittered by the experience of being rejected from dating the shroud by the Archbishop of Turin, Cardinal Ballestrero, or rather his science advisor Luigi Gonella.

When the data finally came in from the two other AMS laboratories that with Arizona had been chosen to engage in this great scientific adventure, Oxford University in England and the Technical University in Zurich, Switzerland the resulting final average was AD 1325 with an uncertainty of ±33 years. A paper presenting the results and authored by 21 people divided among the three institutions and the British Museum was published in the 16 February 1989 issue of the British journal *Nature* [5]. People from the British Museum were co-authors because, as mentioned previously, that institution had played a neutral role in the interlaboratory comparison and later, when the three laboratories had completed their work, in analysing their data.

Although a public triumph for AMS, the revolutionary carbon dating technique first developed at the University of Rochester almost 11 years before, the result was a great disappointment to many who had hoped and in many cases passionately believed that the Turin Shroud was indeed Christ's burial cloth. It is interesting, however, that 'true believers' in the main, continue to believe, hoping that some mistake was made in the radiocarbon measurement. However, no mistake was made.

NATURE VOL. 337 16 FEBRUARY 1989 ————————————ARTICLES———————————— 611

# Radiocarbon dating of the Shroud of Turin

P. E. Damon[*], D. J. Donahue[*], B. H. Gore[*], A. L. Hatheway[*], A. J. T. Jull[*],
T. W. Linick[*], P. J. Sercel[*], L. J. Toolin[*], C. R. Bronk[‡], E. T. Hall[‡],
R. E. M. Hedges[‡], R. Housley[‡], I. A. Law[‡], C. Perry[‡], G. Bonani[§], S. Trumbore[*||],
W. Woelfli[§], J. C. Ambers[¶], S. G. E. Bowman[¶], M. N. Leese[¶] & M. S. Tite[¶]

[*] Department of Geosciences, † Department of Physics, University of Arizona, Tucson, Arizona 85721, USA
[‡] Research Laboratory for Archaeology and History of Art, University of Oxford, Oxford, OX1 3QJ, UK
[§] Institut für Mittelenergiephysik, ETH-Hönggerberg, CH-8093 Zürich, Switzerland
[||] Lamont-Doherty Geological Observatory, Columbia University, Palisades, New York 10964, USA
[¶] Research Laboratory, British Museum, London, WC1B 3DG, UK

*Very small samples from the Shroud of Turin have been dated by accelerator mass spectrometry in laboratories at Arizona, Oxford and Zurich. As controls, three samples whose ages had been determined independently were also dated. The results provide conclusive evidence that the linen of the Shroud of Turin is mediaeval.*

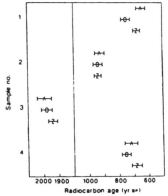

Fig. 1  Mean radiocarbon dates, with ±1σ errors, of the Shroud of Turin and control samples, as supplied by the three laboratories (A, Arizona; O, Oxford; Z, Zurich) (See also Table 2.) The shroud is sample 1, and the three controls are samples 2-4. Note the break in age scale. Ages are given in yr BP (years before 1950). The age of the shroud is obtained as AD 1260-1390, with at least 95% confidence.

*Article in 16 February 1989 issue of* Nature. *The shroud dates to AD 1260–1390 at a 95% confidence level. (Reprinted with permission from* Nature **337** *(1989) 302, copyright Macmillan Magazines Ltd and from P E Damon and D J Donahue.)*

Fears that one of the laboratories might produce another disastrous outlier were, fortunately, not realized. One outlier among three cannot be statistically eliminated as it could among six or seven. The laboratory intercomparison paper presented at the Trondheim Radiocarbon Conference and later published in 1986 in volume 28 of the journal *Radiocarbon* [2] demonstrated how it was possible to detect one outlier among six measurements on well understood statistical principles. If only three measurements are made and one is in substantial statistical disagreement with the other two statistical prudence dictates that all three must be averaged to obtain the final result. However, in this case the agreement among the three for the age of the shroud and of the three control samples was statistically unexceptionable.

Almost as soon as the results were announced, reasons were advanced for doubting or even discrediting them. These included suggestions that the shroud samples were contaminated with organic carbon of a younger age, and I will deal with this later in this chapter as well as charges that the British Museum representative had switched samples during the time he was encapsulating them. This latter procedure was not witnessed by representatives from the three laboratories but at least one other person was present, namely the Archbishop of Turin. The arguably fanciful possibility was also advanced that the resurrection of Christ was accompanied by a burst of neutrons which produced additional carbon-14 thus decreasing the shroud's apparent age. I published an analysis of these arguments for doubting the dating results and an assessment of the results themselves in 1990 in volume 32 of the journal *Radiocarbon* [6].

To those who continue to insist that the shroud must date to the first century AD, the possibility of organic contamination is most often advanced. I had already pointed out in an article published in 1989 in volume 31 of *Radiocarbon* [4] that the samples were taken from the same place on the shroud and that all three laboratories would probably use the same cleaning procedures. This would mean that any contamination of the area not removed by such cleaning procedures would equally affect all three measurements making them in agreement but wrong.

In the event, however, rather different cleaning procedures were employed by the three laboratories, ranging from ultrasonic cleaning to much more rigorous and thorough cleaning techniques. Experts in the field do not know of any contamination that would not be removed by one or other of the cleaning procedures employed. Probably never in the entire history of carbon dating has such care been taken in the measurements. Except in similarly unusual circumstances, it probably never will again.

For those who wish to believe that the Turin Shroud was Christ's burial cloth, despite the radiocarbon dating that gave its age as AD 1260–1390 with at least a 95% confidence, the most commonly invoked reason is that biological contamination changed its radiocarbon age. It is therefore useful to calculate how much contamination would be required to change the measured date from the time of Christ's death (c. AD 0) to AD 1325 (the mean of the above quoted limits) as a function of the time in which the contamination occurred. One will then see how much biological soiling from whatever source would be needed if it had occurred, for example, at the time the shroud's existence was first revealed c. AD 1353 or at the time of the fire in AD 1532 that produced the extensive damage that required the application of the many patches or from recent contamination.

Such a calculation is relatively straightforward and produces the following results: if the biological contamination occurred recently, 64% of the shroud samples supplied to the three measurement laboratories would be contamination and only 36% shroud; if it occurred as a result of the fire in 1532, 86% of the samples would be contamination and only 14% shroud; if it occurred when the shroud's existence was first revealed in 1353, 98% would have to be contamination and only 2% shroud and if it occurred in AD 1325, the date carbon dating established, the sample would have to be total contamination with no AD 0 cloth at all.

It should be noted that these calculations do not take into account the dendrochronological corrections to radiocarbon ages. Such corrections are readily applied, however, and do not seriously affect the results. For example a biological sample that died in AD 1532 would have a radiocarbon age of about AD 1644. That would change the above 86% to 79%, i.e. the shroud sample would still have to be 79% contamination. The dendrochronological corrections for the fourteenth century dates are even smaller.

A photograph of part of the shroud sample received by the Arizona AMS laboratory before cleaning was sent to me, and I also saw the sample itself during the carbon dating in Arizona in May of 1988. Contamination of the magnitude listed above is patently not present. Those who continue to invoke contamination as a possible rationale that the radiocarbon dating of the shroud is in error by 1300 years or so are deluding themselves. When they make public pronouncements to this effect they are committing a disservice to the public.

Charges that the British Museum official who coordinated the radiocarbon measurements substituted fourteenth century cloth for the shroud sample can be dismissed out of hand. Even had he been willing and able to do so, the three to one herringbone twill weave of the shroud could not be matched, as the British Museum officials discovered when they attempted to find such for control samples It was in fact fortunate that the shroud samples were identifiable, so as to preclude the possibility of substitution. Had the measurements been performed 'blind' there would have been no way of knowing that shroud samples were indeed being measured.

There is no way of completely ruling out the possibility that the three laboratories discussed the results of their measurements amongst themselves and made appropriate adjustments to the data before releasing it to the British Museum and, in particular, that they conspired to produce a fourteenth century date for the shroud. All the senior members of the three laboratories have been long-standing acquaintances and, indeed, friends of mine for many years. Although this friendship with some of them cooled somewhat when they 'broke ranks' and submitted to Gonella's wishes, it is completely inconceivable that they connived in any way. As for a conspiracy to produce a fourteenth century date for the shroud, one can only note that such a result was not the one desired by at least one senior member of one of the laboratories. He and his wife are members of the Roman Catholic Church and I am sure they wished for a different result as did many of us involved in the shroud dating enterprise, including me.

One of the arguably more fanciful possibilities for the radiocarbon date of the shroud being in error has been advanced by T J Phillips of the High Energy Laboratory of Harvard University [7]. He pointed out that the resurrection of Christ as described in the Bible was a unique physical event, not accessible to direct scientific scrutiny. One cannot quarrel with that statement. He went on to argue that since the image on the shroud appears to be a scorch, it suggests that the body on it radiated light and/or heat. From this he suggests the possibility that 'it may also have radiated neutrons, which would have irradiated the shroud and changed some of the nuclei to different isotopes by neutron capture. In particular, some carbon-14 could have been generated from (the stable) carbon-13'. This could have changed its apparent age from first century to fourteenth century AD. Phillips advances no theories as to how any biological process can produce intense heat or light, much less neutrons, even thermal ones. To produce the latter requires energies of millions of electron volts (the energy an electron acquires on being accelerated through a potential of 1 volt) whereas biological processes involve energies a factor of $10^9$ less. He does not remark on the astonishing coincidence between the date this ghostly neutron irradiation produced and the known historical date of the shroud c. AD 1353, nor does he note that, since the neutron intensity from such a hypothetical source must vary with distance from the source, the piece of cloth selected was at just the correct distance to produce the historical date. A piece closer to the image, on the Phillips' hypothesis, would have produced an even more recent date. Had such a result been obtained from the AMS radiocarbon measurement, it would certainly have strengthened Phillips' speculation, but would have astonished most of the rest of the scientific community. An elegant answer to Phillips was provided by Robert Hedges [8], a member of the Oxford AMS laboratory involved in dating the shroud, who pointed out, among other things, that if one wanted to invoke 'miracles' of the kind proposed by Phillips then there was no point in any scientific investigation of the shroud at all.

More recently a Russian investigator, Dimitri Kouznetsov, and his collaborators claim to have established experimentally that heating a linen cloth in a humid atmosphere rich in carbon monoxide and carbon dioxide augments the amount of both carbon-13 and carbon-14 in the cloth [9]. They argue that this happened in the fire of 1532 and that the increase in carbon-14 caused the carbon date to change from its real value of first century to an apparent value of fourteenth century. In this connection it should be noted that, prior to the meeting in Turin in 1986 to establish procedures for dating the shroud, tests had been carried out at one of the AMS laboratories to determine whether heating linen in air to the point of charring it changed the ratio of carbon-14 to stable carbon. The answer was that it did not. That, plus the fact that the carbon-13 to carbon-12 ratio measured in the shroud linen by the three AMS laboratories was normal, makes Kouznetsov *et al*'s claims highly suspect. In the same journal that contained the Russian paper an answer [10] was provided by the three senior scientists at the AMS laboratory at the University of Arizona who carried out the first carbon dating of the shroud [5]. They duplicated the experiments carried out by the Russian group and concluded that the results claimed by Kouznetsov and his co-workers were unverifiable and irreproducible.

There is one other development that deserves further investigation. All the reasons discussed above for the shroud dating to the first rather than to the fourteenth century and others not mentioned range from the highly improbable to the ludicrous. The one to now be discussed needs to be taken more seriously. It involves bacterial and fungal infestation of ancient cloth. That possibility was first suggested by Leoncio A Garza-Valdes, a paediatrician and bacteriologist in San Antonio, Texas.

In September 1994 a round table on the microbiology of ancient artefacts was organized by Garza-Valdes and Stephen Mattingly, a microbiologist at the University of Texas Health Science Center at San Antonio (UTHSCSA). Garza-Valdes also holds a visiting appointment at UTHSCSA. The subject of the round table concerned bioplastic coatings produced by bacteria and fungi and found on the surface of ancient artefacts, desert rocks (where it is referred to as desert varnish) and around the fibres of many ancient textiles including the Turin Shroud. These bacteria probably take their nourishment from the air and hence can be adding carbon with a component of carbon-14 contemporaneous with the time of bacterial growth—and thus having a higher ratio of carbon-14 to stable carbon than that of the carbon in the cellulose of the ancient textile.

With the assistance of Giovanni Riggi (the person who removed a sample from the shroud in 1988 for radiocarbon dating) Garza-Valdes obtained a small sample of the shroud taken from the same area as those used in the AMS measurements. Microscopic examination showed a definite halo or bioplastic coating of varying thickness around the fibres. The UTHSCSA researchers established that the acid–base–acid cleaning procedure employed on the shroud samples by the three laboratories left the bioplastic coating intact. They concluded that the effect of this coating might have caused the carbon date to be too young. How much too young they could not estimate because a procedure for separately carbon dating the bioplastic coating and the cellulose from ancient linen has not yet been demonstrated. It is worth noting that because these bacterial infestations form surficial coatings, if they affect the radiocarbon date at all they are most likely to have their maximum impact on the radiocarbon measurements of the ages of cloth as opposed, for example, to parchment. The surface to volume ratio of cloth vastly exceeds that of other organic artefacts.

Support for the possible influence of bacteria on radiocarbon ages of cloth comes from radiocarbon studies of Egyptian mummies. There is some evidence that the wrappings of some of these mummies yield radiocarbon dates that are younger than the bones of the mummies. The collagen from the bones should

*A statue of a sacred ibis.*

not have suffered any bacterial contamination and should yield the
true date on which the mummified person died. In particular the
Manchester mummy 1770 yielded a linen wrapping date that was
$340 \pm 147$ years younger than the collagen from human mummy
bone [11]. Whether this was a result of a re-wrapping of the
mummy, poor preparation of the collagen sample or bacterial
contamination is not known.

One way to obtain information on the impact bacteria may
have on the radiocarbon ages of ancient textiles is to investigate
non-human mummies where the odds of re-wrapping are low.
After all, those humans who were mummified by the Egyptians
must have had considerable status while alive. If their wrappings
began to deteriorate or to become shoddy in appearance at some
later date it would only be natural to re-wrap them with cloth that
was now younger than the bones.

The Egyptians mummified a variety of animals and birds
including cats, bulls and ibises. The presence of bioplastic coatings
on the threads of the linen wrappings of a mummified Egyptian
sacred ibis in the possession of Garza-Valdes was established. (See
the colour section.) The ascribed age of the ibis was between 30
and 330 BC. Samples of the cloth of the wrapping and bone and
tissue samples of bird were removed for carbon dating. The results
showed that the bone and tissue samples dated to about 800 BC
and that the cloth was some 400 to 700 years younger [12].

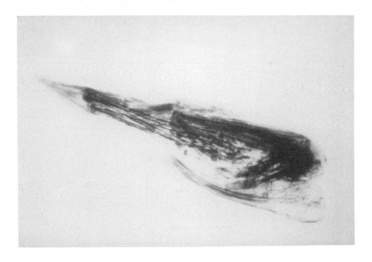

*X-ray of the ibis mummy. Courtesy of L A Garza-Valdes.*

Although the provenance of this ibis is unknown it probably was bred and raised at some temple site along the Nile. Pilgrims to these temples would purchase them and they would be sacrificed and mummified at the site as an offering to the gods. They would then be stored in vast underground galleries at the temple site. Hundreds of such mummified ibises have been found. The likelihood that any of them were re-wrapped, although possible, seems remote.

It was pointed out, however, that if the diet of the ibis in question was marine in origin, for example fish or molluscs coming from the Mediterranean or the Red Sea, the carbon-14 content of the carbon in the bird's body would be lowered and the bird's age would appear to be older. The reason for this is that, because of mixing in the oceans, the average carbon-14 to stable carbon ratio is lower than the contemporary terrestrial value. It seems unlikely that the ibises raised in captivity in temples along the Nile were fed marine food from bodies of water a hundred miles or more away but one cannot totally discount that possibility. Because of this the choice of the ibis for this test was not the most propitious. Research on the possible effect of bacterial infestation on the carbon-14 age of ancient textiles is continuing.

For the time being it would seem logical to conclude that the linen from which the Turin Shroud was woven was harvested in the fourteenth century. It most likely was not the burial cloth of

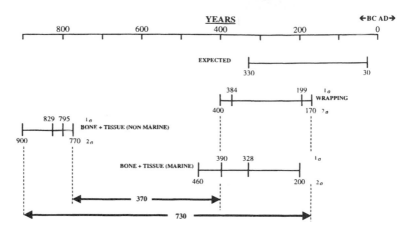

*Results of the AMS measurements on the ibis mummy [12]. The expected age according to A R David is somewhere between 30 and 330 BC. If the ibis ate non-marine food the cloth is somewhere between 370 and 730 years younger than the ibis bone. If the ibis had a marine diet the bone and cloth ages agree and agree with the ascribed age. Both 1σ and 2σ values for the AMS dates are given (see the discussion on the Dead Sea scrolls in Chapter 11 for the definitions of 1σ and 2σ).*

Christ. High priority should be given to the job of preservation of what is still a very beautiful and historic object. Medieval art experts and experts in related scientific areas should be allowed by the Turin ecclesiastical authorities to proceed to determine how the image of a crucified man depicted on the shroud was produced and, if possible, the name of the artistic genius who produced it.

This abbreviated account of the dating of the Turin Shroud omits many rich and fascinating details. It was the most complex 'scientific' endeavour with which I have been involved in my fifty years as a scientist and it occupied a fair fraction of my time for some ten of those years—a vastly longer amount of time than I should have spent. It would have been much more interesting had the shroud's age dated to the first century. That, of course, would not have proved it was Christ's shroud but it would have upped the odds considerably. However, as a scientist, I have to live with what, to many scientists and non-scientists alike, is a disappointing result of very careful scientific measurements.

The shroud will soon again lie in its silver cask in the Cathedral of John the Baptist in Turin. A new archbishop presides over the

Turin archdiocese. No further scientific forays on its most famous artefact have been authorized. The mystery of how its image was produced remains just that—a mystery. Maybe that is how it should be.

# Chapter 11

# The Iceman, the Dead Sea Scrolls and More

A fascinating aspect of accelerator mass spectrometry is the remarkable variety of scientific fields to which it can be applied. Many of these have been covered in previous chapters. Not all the world's AMS laboratories work in the same scientific areas but a number of them measure, primarily, carbon-14 to establish the ages of ancient artefacts. To the general reader this is probably the most interesting application. The original motivation for the development of AMS was to accomplish the direct detection of carbon-14 in organic samples. This permitted the dating of samples at least a thousand times smaller than was ever previously possible by decay counting and opened up new vistas in the fields of archaeology and anthropology. Other laboratories have more catholic interests and include other radioactive and stable isotopes in their repertoire.

This chapter describes a somewhat mixed bag of interesting AMS applications not already covered. Included will be AMS studies of Ötzi, the Neolithic 'Iceman', the egg of the elephant bird, desert rat urine, extending the carbon-14 time-scale calibration back to 30 000 years or more, applications of AMS in biomedicine, measurements of the age of old ground water, the exposure ages of rock surfaces and the ages of the Dead Sea scrolls.

The most spectacular item in the above list is the discovery and subsequent carbon dating of the Neolithic Iceman, nicknamed Ötzi after the Ötztal Alps on the border between Austria and Italy where the body was found. On 19 September 1991 two German hikers discovered his frozen corpse in the Similaun glacier

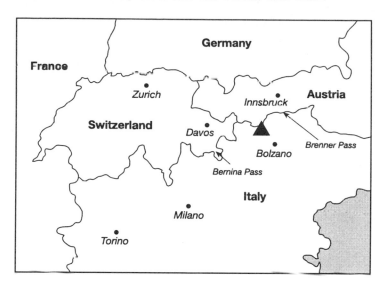

*A map of the alpine regions of Austria, Switzerland and Italy showing (▲) where
the Neolithic Iceman was discovered in 1991 in the Ötztal Alps on the border
between Austria and Italy.*

in the Ötztal Alps 10 500 feet above sea level and, as was later
ascertained, about 300 feet inside the Italian border. Ötzi's body
resembled the starved and shrunken cadavers found in the Nazi
death camps at the end of World War II. During the first four days
after the discovery a number of well-meaning people tried to hack
the body out of the ice using ski poles and ice axes. Considerable
damage to the Iceman's clothing, to some of his equipment and
to the body itself resulted from their rather clumsy and inept
attempts at recovery. When it was finally freed from its icy tomb
the body, still frozen solid, and some of the implements found
with it, including a bronze or copper axe, were transported to
Innsbruck, Austria, by plane and car. (See the colour section.) A
day later the body was first seen by Dr Konrad Spindler in the
morgue of the University of Innsbruck. To Spindler the shape of
the axe blade resembled that of relics of the early Bronze Age so
he immediately guessed that the Iceman died some 4000 years ago
and not 500 years as immediately speculated in the press.

Professor Spindler, of the Research Institute for Alpine Studies
at the University of Innsbruck, is currently leading an international
team studying Ötzi and the artefacts found with him. I met him
in Canberra, Australia at the Sixth International Conference on
Accelerator Mass Spectrometry held in Canberra and Sydney from

27 September to 1 October 1993 (see reference [9], Chapter 3). He had been invited by the conference organizers to give a public lecture on Ötzi on 28 September at the Australian National University. He was accompanied by his colleague, Dr Romana Prinoth-Fornwagner, who is also a member of the Alpine Studies Institute and who is analysing the carbon-14 data from other circumalpine Neolithic sites to try to associate Ötzi with specific Neolithic cultures.

Spindler told me that an article on the Iceman had appeared in the June 1993 edition of the *National Geographic* [1]. He was not enamoured of the treatment the author had given to the subject. It is, none the less, an interesting account containing, as is part of *National Geographic*'s well deserved reputation, many excellent photographs. For Spindler's taste there was far too much speculation in the article on where the Iceman came from, how he reached his final resting place, what he was doing there, how he died and what life in the early part of the Copper Age (4000–2200 BC) was like. A much more factual account appeared in the February 1993 edition of *Popular Science* and a year later in the March 1994 edition of the same magazine.

Spindler was to deliver his address in the Manning Clark Theatre at the Australian National University in Canberra and the hall was packed. His talk was scheduled to last an hour and it was to be followed by a half hour talk by his associate, Prinoth-Fornwagner, who would discuss carbon dating. In his talk, Spindler excused his English which was, in fact, excellent. He said it was his first visit to Australia. He noted that the metal of the axe blade he had first assumed to be bronze was actually copper. The Copper Age had lasted up to about 2200 BC and thus the Iceman's age was probably even older than he had at first estimated. Spindler stated that the Alps of this period were rugged and wild but were, by no means, an impenetrable barrier to prehistoric man. Hardy travellers of the Copper Age passed through on a regular basis on trading and hunting expeditions. Ötzi may well have been a Copper Age trader.

The principal items found with the body included a 1.82 metre long incomplete bow made of yew but broken during the body's recovery, eleven 85 centimetre long arrows—two of them tipped with double-faced flint points—the world's oldest quiver, made of deer skin, a 10 centimetre copper axe with its bindings and handle intact, an ash-handled flint dagger with its grass sheaf, a bone needle and a 25 centimetre stick tipped with an antler. Of these the most remarkable was the axe. It was cast from molten metal poured into a mould. After it cooled, it was further shaped with a hammer. The blade was inserted into an L-shaped crook of yew wood and lashed in place with leather thongs soaked in glue, possibly made from birch tree sap. The broken bow bears evidence of some initial shaping possibly carried out using the axe.

The Iceman was 25 to 35 years old when he died. When alive, he was five feet three inches tall and weighed some 110 pounds. His corpse now weighs 44 pounds. He had dark hair and a beard. His lower back bears the marks of short parallel tattoos. His clothing, stripped off during the bungled rescue procedures, comprised carefully stitched tanned deer skin, a skilfully woven grass cape that may have doubled for a bed roll, puttee-like leather leggings and expandable leather shoes with multiple eyelets that permitted the shoes to be stuffed with wads of insulating hay. When the body reached Innsbruck it was totally naked except for one hay-stuffed shoe. Much of the remaining clothing was recovered later and is now being carefully pieced together at the Römisch-Germanisches Zentralmuseum in Mainz, Germany. All of the other artefacts found with the body are in Mainz as well. At present, the body itself, swathed in a cocoon of ice and plastic that maintains a surface humidity of about 98%, is lying in a freezer in Innsbruck at a temperature of −6 °C (21 °F) and can only be studied by scientists for 20 minutes every two weeks.

The Alpine Iceman constitutes one of the major anthropological discoveries of the century. As such it was of paramount importance to establish its age. Carbon dating was clearly the way to do so. However, before the invention of AMS, that would have posed a serious problem. An unacceptably large sample of his flesh and bones would have to be converted into $CO_2$ for radioactive decay counting by the Libby method exclusively employed up until 1977. However, AMS was by then and had been for many years a mature science and could be safely employed to date Ötzi.

Following Professor Spindler's talk, Dr Romana Prinoth-Fornwagner discussed the preliminary AMS carbon-14 dating results. Her accent was Italian—in fact she claimed she spoke English like the pope! A number of European AMS carbon dating laboratories were involved, including Gif-sur-Yvette, Oxford,

Uppsala and Zurich. Oxford and Zurich received bone and tissue samples and their measurements were in excellent agreement. However, because of wiggles in the carbon-14 dendrochronology calibration curve the age of the Iceman could not be pinned down with the accuracy inherent in AMS. The best one can say is that he died between 3110 and 3370 BC or some 5200 years ago. In any case Ötzi is the oldest and best preserved human body ever found. He outdates King Tutankhamun by almost 2000 years. The botanical material was also dated by AMS. Most of the carbon dating results have now been published in the Canberra AMS conference proceedings by Prinoth-Fornwagner and Niklaus [2].

In the same conference proceedings there is an article by Spindler [3] in which he weaves all the data together to provide a fascinating account of the Iceman's last weeks. He suggests that Ötzi may have been chased out of a lowland settlement and pursued into the mountains at the onset of winter, a journey he would hardly have undertaken willingly at that time of year. The season he fled was established by the fact that he was carrying berries that ripen in late fall. Pollen and kernels of grain were also found on his clothing, suggesting he came from a valley farming community. Three of the ribs on the right side of his body had been freshly broken shortly before he died. Such injuries could have resulted from an altercation in the village before he fled. Spindler has also speculated on the purpose of the tattoos found on the lower back of the Iceman's body. X-ray and other measurements indicate that the Iceman had suffered many injuries throughout his life. His nose had been broken several times and ribs on his left side had been fractured earlier in his life. There was also evidence that he suffered from a bad back. Some Neolithic 'doctor' might have treated the condition by branding the skin above the aching area and then rubbing herbs into the wounds. This treatment, as is sometimes the case even today, seems worse than the condition it was supposed to correct.

One of Dr Prinoth-Fornwagner's tasks was to try to identify Ötzi's home territory, and on that question she had reached a conclusion. She ended her talk by saying, 'Since today I am far from Europe I dare to tell you that the Iceman was an Italian'. She added that her children wondered when Steven Spielberg would make a movie about the life and times of Ötzi.

While in Australia I visited the Australian National Tandem for Applied Research (ANTARES) facility at Lucas Heights outside

Sydney. This laboratory houses an FN tandem (an accelerator with a terminal voltage up to about 8 MV) that formerly belonged to Rutgers University, New Jersey, USA. It was purchased by the Australian Nuclear Science and Technology Organization (Ansto) at Lucas Heights, of which ANTARES is a part, in 1989, after the accelerator was shut down. For some 25 years it had been used at Rutgers for nuclear physics research. Ansto is a very large Australian national laboratory akin to the national laboratories in the USA at Brookhaven, Oak Ridge, Argonne, Los Alamos, Berkeley and Livermore. ANTARES is a new and a relatively small part of the main laboratory and, in some sense, is struggling for recognition in Ansto's larger scheme of things. About 50% of the beam time available at ANTARES is devoted to AMS and of this, at present, carbon-14 measurements are dominant. It was during this visit I learned the fascinating story of the egg of the ancient Madagascan elephant bird.

Some time in the spring of 1993 three children were playing in the sand dunes at Cervantes about 150 miles north of Perth in Western Australia. They found a giant coffee-coloured egg buried in the sand. The egg was 31.7 inches in circumference and had a capacity of almost 2 gallons—the equivalent in size to 150 hens' eggs. The find was brought to the attention of John Bell, a Perth auctioneer, who realized it might have considerable value and the children's parents suggested that he might auction it off with the proceeds going to the children who found it. It might be worth up to 70 000 Australian dollars, Bell speculated. He may have envisioned an auction at Sotheby's in London where the eccentricities of wealthy British collectors would push the price up. Again the key questions were how old was the egg and what species of bird had laid it? Bell had heard about Ansto's carbon dating expertise from museums in Western Australia and so, in May 1993, he brought it to Dr Claudio Tuniz, head of ANTARES, to find out whether his AMS laboratory could date the egg. Indeed Claudio and his AMS colleagues could.

Egg shells are mainly composed of calcium carbonate and the carbon in this compound has, as its origin through the food chain, the carbon dioxide gas present in the atmosphere at the time the egg was laid. One part in a trillion of the carbon in this gas, in the food the bird ate and thus in the egg shell would have some carbon-14 as a small component in place of stable carbon. The carbon-14 in the shell would decay without replenishment as soon as the egg was laid. Measuring the amount of carbon-14 compared

*The egg of the Madagascan elephant bird alongside a pack of cigarettes. The egg is approximately 13 inches (33 centimetres) high and 10 inches (25 centimetres) in diameter at its middle. It has a volume of almost 2 gallons or the equivalent of 150 hens' eggs.*

with stable carbon would reveal how long ago that happened. Since the researchers at ANTARES would be using AMS for the carbon-14 measurement they would need to remove such a small amount of the shell that its loss would never be noticed. Again, the measurement of carbon-14 in such an exquisitely small amount of shell—less than a milligram or so—would not have been possible with the pre-1977 carbon dating method without destroying an unacceptably large amount of egg. The ANTARES measurement showed the egg was laid $1970 \pm 73$ years ago. According to Dr Tuniz that age for the egg made it absolutely certain that it was not produced from an Australian prehistoric giant emu.

It was soon established, with little uncertainty, that the egg came from the Madagascan *Aepyornis maximus*, a huge flightless emu-like bird, with the common name elephant bird. It was first recorded on the island of Madagascar in 1505 and was last seen in 1808. The bird weighed close to 1000 pounds and was 8 feet tall. How the egg managed to travel undamaged some 5000 miles from Madagascar to Western Australia is a mystery. Could it have floated all that distance across the Indian Ocean? Dr Patricia Rich of Monash University in Melbourne said she doubted it did because there was no evidence of the egg having been covered with marine organisms. She considered it more likely that it came off a ship-wreck since such eggs were traded as curiosity items in the nineteenth century. There are ten specimens of elephant bird eggs in the world and it is the largest egg known.

At this point, some time in September 1993, the Western Australian government stepped in. They claimed that, because it was found on Crown land, it belonged to the government. The egg was then still in the possession of the three children, and rather than have it snatched from their hands by officialdom they went back to the dunes and reburied the egg. It was the sort of cheeky action that must have been applauded by many Australians. The Western Australian government then had second thoughts and agreed to pay the children 73 000 Australian dollars of which 11 000 dollars would come from the government and the rest as a result of a public appeal after the egg was put on display in the Western Australian Museum in Perth. The children dug up the egg and handed it over to the museum. It was insured for 114 000 dollars.

Whatever the final amount of money the children received, the incident and the favourable publicity it engendered reflected credibly on the ANTARES laboratory and on its director Claudio Tuniz in the eyes of the top brass of Ansto. It did much to establish ANTARES' place as an important unit in the much larger national laboratory. One hopes that it also made government officials have second thoughts about how to deal with young people.

The next item, at first sight, seems a trifle bizarre. It concerns the urine, faeces and other material found in the middens of desert wood rats. My first introduction to the scientific potential of the nests of this species of rats of the genus *Neotoma* came in a talk given at the first conference on accelerator mass spectrometry which I and my two AMS colleagues, Ted Litherland of the University of Toronto and Ken Purser of General Ionex Corporation, organized (see reference [1], Chapter 3). It was held at the University of Rochester in April 1978. The speaker was V C LaMarche, Jr a scientist at the Laboratory of Tree Ring Research at the University of Arizona in Tucson [4].

He pointed out that the middens constructed by these rats are widely found throughout the western United States. The animals incorporate a wide range of well preserved vegetal materials, mixed with faeces and urine, into their nests. Because desert rats consume very little liquid water apart from the water in plants their urine is viscous and hardens into a crystalline solid. Where protected from moisture in caves, rock shelters and crevices such deposits can be extremely old—as old as 40 000 years, for example.   They are of considerable importance in establishing ancient vegetation patterns and climates in the regions in which they are found.  Because most of the material in the middens is organic it can be carbon dated.  One approach would be to pick out single seeds or twigs and carbon date them individually.  That would not be feasible using carbon dating techniques available before 1977 but LaMarche recognized the potential of AMS as a carbon dating tool to do the job.  He realized, as he indicated in his talk, 'It should now be possible to date individual midden components in considerable detail, and to unambiguously assign radiocarbon dates to each of the components of the flora that existed locally at different times in the past'.  An obvious problem is that, at the time LaMarche gave his talk, the only way to calibrate the time variations in the cosmogenic production rate of radiocarbon was by its measurement in tree rings residing in the collection of the Tucson laboratory in which LaMarche worked.  Such tree rings only go back some 8000 years.  A way of circumventing this limitation has recently been invented as we will see later.

Radiocarbon studies of the biota in the rat nests have been made by AMS, but an even more exotic exploitation of this rat urine has been carried out, and again AMS is playing a key role.  The results were headlined in the 6 April 1992 issue of *Time* 'Nature's time capsules' and the *New York Times* edition of 26 January 1993 'Ancient desert rats leave clues to kaleidoscope of climate change'.  The *Time* article was written by Leon Jaroff and that in the *New York Times* by one of the paper's crack science writers, Malcolm W Browne.  The research involves the measurement of both carbon-14 and chlorine-36 in crystallized desert rat urine and was conducted by F M Phillips of New Mexico Tech at Soccoro, New Mexico, P Sharma of the University of Rochester and P E Wigand of the Desert Research Institute in Reno, Nevada.  The chlorine-36 in the crystallized urine was produced in the upper atmosphere by cosmic rays, fell to the ground in precipitation and entered the food chain of the rats.  The measurements depended, for their success, on the extension of the calibration for cosmogenic radiocarbon production to times up to almost 30 000 years—almost four times longer than the dendrochronological (tree ring) limit.   Measurements of radiocarbon in tree rings was described in Chapter 2.  How was this extension accomplished?  Counting tree rings provided an

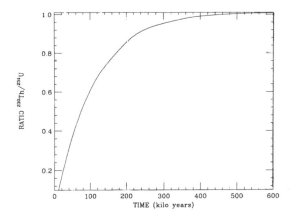

*Uranium-234 and thorium-230 are two elements that occur in the decay chain that is initiated by uranium-238. The $^{230}Th/^{234}U$ ratio increases with time over a period of almost half a million years. The measurement of this ratio in coral, for example, can extend the calibration curve for corrections to carbon-14 dates from the present 10 000 years, using tree rings, to 60 000 years or more.*

independent clock for calibrating the cosmogenic production of carbon-14 but it only goes back in time some 8000 years. Can a clock be found that goes back further?

As mentioned Chapter 9, Ernest Rutherford was awarded the Nobel Prize for chemistry in 1908 for his studies of the decay chains of uranium and thorium. In particular he found that the decay chain of uranium-238 involves a series of alpha and beta transitions, totalling 14 in all, to reach the final stable isotope of lead, lead-206. Uranium-238 is almost stable. It has a half life of 4.5 billion years, approximately that of the age of the earth. This means that half the uranium-238 that was present when the earth was formed is with us today, although that is not relevant to the present discussion.

In the complex decay chain that uranium-238 initiates, the ratio of the amounts of two of the subsequent elements, thorium-230 and uranium-234, increases steadily with time over a span of almost half a million years. Uranium-234 is present in sea water and in any material containing uranium-238 and is incorporated into corals as they grow in the sea. By accurately measuring this thorium-230 to uranium-234 ratio in parts of the coral that have stopped growing, the number of years ago that growth ceased can be determined. Because coral is an organic material it also contains carbon-14. By measuring the thorium-230 to uranium-234 ratio and the carbon-14 to stable carbon ratio in the same piece of coral one can, in principle, construct a calibration curve for radiocarbon dating that extends to the latter's present limit of some 60 000 years.

The measurements were carried out in 1989 and 1990 by scientists from the Lamont-Doherty Geological Observatory at Columbia University in New York. The corals employed were raised off the island of Barbados and the results were published in *Nature* [5]. So far the coral samples have U–Th ages that go back only as far as 30 000 years and at this present limit the radiocarbon ages are too young by close to 3000 years. Twenty thousand years ago the activity of carbon-14 in the atmosphere was about 50% higher than today. It is believed that the results can be explained by fluctuations on a gradual decrease in the earth's magnetic field as one goes back in time. In any case one now has a calibration curve for carbon-14 dating that extends the dendrochronological calibration from some 8000 years to 30 000 years, although not in nearly so much detail as the dendro calibration for the most recent 8000 years.

The study of both chlorine-36 and carbon-14 by Fred Phillips and his collaborators in desert rat urine was designed to answer the question as to whether the cosmogenic production of chlorine-36 varied in a similar way to that of carbon-14, at least back as far as 30 000 years ago. Not too surprisingly they found that it did. It was an important and interesting but hardly earth-shaking result. Why then did it receive such media attention? The answer can only lie in the novelty and mildly repugnant nature of the material that formed the basis of the study—crystallized rat urine. Many readers of the *New York Times* article were probably more intrigued to learn that, 'Several Native American tribes prepare a kind of tea from crystallized rat urine, which they drink as a medicinal cathartic' than that studies of the urine revealed arcane secrets of the past cosmogenic production of chlorine-36.

The variation of the cosmic ray flux incident on the earth provides information on the variation in the earth's magnetic field. A low magnetic field provides less of a shield against cosmic rays and permits an enhancement in their intensity. The increase in the production of both chlorine-36 and carbon-14 some 20 000 years ago compared with today means the earth's magnetic field has strengthened since then. Would it be possible to get similar information even further back in time? The measurement of U–Th ratios in any form of matter provides an absolute clock back as far as half a million years. If the material also contains chlorine-36 with its half-life of around a third of a million years one can learn how the earth's magnetic field has varied for the last half million years. The problem is to find formations that grow with time, are half a million years old and contain both uranium and cosmogenic chlorine. Perhaps stalactites and stalagmites would serve, but that has not yet been demonstrated.

Another topic in this pot-pourri of AMS applications concerns biomedical research. At the First International Conference on Accelerator Mass Spectrometry held at the University of Rochester in April 1978 it was recognized that AMS could play an important role in such research. Two scientists at the university's Medical School presented a paper [6] in which they noted that, for the past 35 years or so, tracer kinetic methods have been widely employed in medical research in studies of human metabolism. It was their feeling that improved techniques that could be employed by the use of AMS in such studies would not only have diagnostic importance but would be of value to pharmacological studies where tracer kinetics are commonly employed.

They gave as a typical example research they had carried out on the conversion of glucose to the amino acid alanine in human subjects. In this study a measured dose of glucose in which stable atoms of carbon at key sites in the glucose molecule had been replaced with carbon-14 was injected into the vein of a subject after an overnight fast. At subsequent times over a period of 3 hours samples of blood were drawn and the glucose and alanine were chemically separated. The amount of carbon 14 in each was measured in a conventional radiocarbon decay counter. The results provided important information on certain metabolic pathways in humans.

The total amount of blood that was extracted over the 3 hour period was about a third of a pint (1 pint is extracted from blood donors by the American Red Cross, and such donations can only be repeated every eight weeks). The samples were taken every 15 minutes during the first hour and every half hour during the second 2 hours. The researchers noted that the data would be much improved if smaller blood samples could be drawn more frequently, say at 3 minute intervals, but this would require unacceptably high doses of carbon-14 tagged glucose to be administered. As it was the amount of carbon-14 that was administered was close to the legal limit.

The experiments were limited by the maximum permissible dosage levels of carbon-14, the frequency with which blood samples can be taken and the blood sample size required to overcome the background limitation of radiocarbon scintillation decay counting. Detection of the carbon-14 by AMS would permit smaller dosages of the labelled glucose and the extraction of smaller blood samples at more frequent intervals. The two researchers noted that the AMS detection technique appeared very attractive but development was needed and cost might be a problem. They concluded, 'But the need is strong, and the alternatives limited, and these circumstances may provide impetus for the necessary development'. They were remarkably prescient.

Surprisingly enough, nine years later, in 1987 no liaison between AMS experts and experts in the biomedical field had yet formed to exploit the power of AMS in medical research. A paper published that year [7] discussed an application of AMS in the biological sciences involving carbon-14 autoradiography.

A possible reason why AMS laboratories specializing in the detection of radiocarbon for the dating of carbonaceous artefacts of organic origin have not leapt into the biomedical field is the fear that chemicals enriched in carbon-14 well above that found in modern organic carbon would contaminate very old samples they have occasionally to measure. But probably a more important reason was the absence of a fortuitous congruence of people in the AMS field and in the biomedical field who were willing and able to speak each other's language. A group at the ·Lawrence Livermore National Laboratory has solved that problem.

The first applications of AMS to the biomedical sciences were reported in 1990 by a group of nuclear physicists and biomedical scientists at the Center for Accelerator Mass Spectrometry of the Lawrence Livermore National Laboratory in California [8]. One paper, presented at the Fifth International Conference on Accelerator Mass Spectrometry held in Paris, France in April 1990, reported the use of AMS to determine the amount of carcinogen bound to animal DNA at levels relevant to human exposure [9, 10]. The results demonstrated the relationship between carcinogen dose and these DNA products in mice that had been given low levels of a complex organic carcinogenic compound found at parts per billion levels in cooked hamburger meat. The carcinogenic compound in question was labelled with a single carbon-14 atom at the point where the carbon connected to an amino acid group. One day after being fed the carcinogen the animals were, as the researchers euphemistically described it, 'sacrificed' [10], the livers were removed and the liver DNA was extracted. The DNA was diluted appropriately and was then converted to graphite. The graphite was inserted into the ion source of the AMS tandem accelerator facility at the Lawrence Livermore laboratory and the carbon-14 to stable carbon ratio was measured.

The results obtained were at least an order of magnitude improvement over the best sensitivity offered to date by the post-labelling assay method involving a radioisotope of phosphorus and three to five orders of magnitude better than other techniques used for quantitative assays of DNA products. They stated that the main limitation of the technique was radiocarbon contamination in their medical laboratories with a long history of carbon-14 use. At that time they expected that up to a ten-fold increase in sensitivity could be achieved through the reduction in contamination.

The Livermore group noted [10] that, 'Clinical applications and research with human subjects can be envisioned with AMS radioisotope tracing. The detection sensitivity and small sample size requirements of AMS make it ideal for measurements of small quantities of easily accessible human cells, in addition to the liver tissue demonstrated [in the experiments outlined above]. Therapeutic parameters for individuals could be determined by AMS through administration of small doses of radiocarbon-labelled pharmaceuticals. Such custom tailoring of effective therapeutic regimens would be particularly valuable for cancer chemotherapy as the extremely small human radiation dose from the drug should not be an issue.' They pointed out that the effective radiation dose involved in their study corresponded to about 0.1% of the total annual adult exposure to ionizing radiation from known natural sources.

At this same meeting a study was reported [11] by researchers from Purdue University and Argonne National Laboratory on the use of calcium-41 as a long-term biological tracer for bone resorption. In their paper these researchers noted that, 'Absorbtion and metabolism of calcium is of great interest because of the widespread incidence of the metabolic bone disease osteoporosis and the effects of space travel on the skeletal system. Osteoporosis manifests itself by a general loss of bone mass. It afflicts 20 million in the USA, or nearly 10% of the population. It can start early in life and lead to fractures that are disabling or fatal later in life (30% of women over 90 have had hip fractures). Similarly, bone loss has been reported by American and Soviet space programs and is thought to be induced by the lack of gravity on the skeletal systems, similar to prolonged bed rest.'

In the experiments, mature beagles were injected, over a 4 week period, with two radioisotopes of calcium, calcium-45 with a half-life of 163 days and calcium-41 with a half-life of 100 000 years.   After some 26 weeks from the time of injection both calcium radioisotopes were measured in the dog's serum, calcium-45 by decay counting and calcium-41 on the AMS facility at the University of Pennsylvania.   The results demonstrated that both radioisotopes of calcium behave identically *in vivo* but that calcium-41 has a sensitivity 100 times higher than calcium-45. Furthermore the advantage of calcium-41 is that, even with a negligible radiation dose, it can be measured by AMS long after the calcium-45 becomes unmeasurable.   The latter radioisotope of calcium is fairly widely used in biomedical research where it is detected by decay counting but it seems clear from the above and other studies employing calcium-41 and AMS that this will become the method of choice in the future.

As reported at the Sixth International Conference on Accelerator Mass Spectrometry held in Canberra and Sydney, Australia in the fall of 1993 a pilot experiment on a human subject is being carried out, again using the radioisotope calcium-41 [12]. A quantity of this radioisotope amounting to a dose that was a factor of 30 000 times smaller than the dose from natural radiation was injected intravenously into a medical doctor in British Columbia, Canada in early 1991 and urine samples have been taken as a function of time over a period of about 2.5 years since then. The calcium-41 to stable calcium ratio, in about a year, reached an equilibrium value.  It is being measured at different points in the menstrual cycle and through menopause. The AMS measurements are being carried out at the tandem accelerator facility at Rehovot, Israel.

Aluminium is the third most abundant element in the earth's crust and can be found in many foods, drugs, plants and drinking water.   It has been identified as an important toxin especially in patients with renal failure, and has also been suggested as a contributor to Alzheimer's disease.   *In vivo* studies of the biological effects of aluminium on animals and humans can be carried out using the radioisotope aluminium-26 with a half-life of 750 000 years.   It is readily detected in exquisitely small amounts by AMS. Preliminary studies of aluminium kinetics in rats have been carried out using the AMS facility at the University of Pennsylvania [13]. Another study of the biological and toxicological applications of aluminium-26 has been carried out by AMS using the very large tandem accelerator nuclear structure facility at the Science and Engineering Research Council Laboratory in Daresbury, UK [14]. In this work aluminium-26 was administered orally to a human subject. Blood samples were taken and the presence of the isotope in one particular blood protein was unambiguously demonstrated as was its absence in other blood proteins. Aluminium-26 was also used as a tracer in experiments to study the uptake of aluminium by human neuroblastoma cells in culture (in relation to the cause of Alzheimer's disease) and in investigations of aluminium speciation in natural water (in relation to the toxicity of aqueous aluminium in fish). This research has, regrettably, been terminated owing to the UK government-imposed shut down of the Daresbury tandem.

The aluminium-26 studies on rats have been extended to normal human subjects [15] with the AMS measurements being carried out on the large tandem accelerator in Rehovot, Israel. One person was injected with a dose of aluminium-26 that constituted less than 0.1% of the natural yearly radiation dose. The removal of aluminium from the body followed a time course that is very similar to the earlier studies on rats. The results establish a baseline behaviour for normal subjects and are a prelude to measurements on individuals who exhibit abnormal ability to clear aluminium from their system. In another study [16] aluminium-26 was administered orally to five healthy human volunteers in the presence and absence of silicon to test the potential for silicon to reduce the bioavailability of aluminium. The results indicated that dissolved silicon is an important factor in limiting the absorption of dietary aluminium.

There are many more biomedical applications of aluminium-26 that are made possible by AMS measurements. A review of such studies has been published recently by Flarend and Elmore [17]. In that article they conclude "The use of aluminum-26 as an isotopic tracer for aluminum has now been demonstrated in a variety of human, animal, plant and *in vitro* systems. These tracer studies have demonstrated the variety of questions that have been successfully addressed in pharmacokinetic, bioavailability, metabolism and toxicological applications. AMS has been proven to be a very effective instrument for measuring aluminum-26 in extremely low concentrations. This has allowed the tracer to be used for studying the kinetics of physiological concentrations of aluminum within single cells and over an extraordinary concentration range of chemical speciation of aluminum. Aluminum-26 can now be used to address questions concerning the improved use of aluminum-containing drugs, mechanisms by which aluminum can be a toxin, bioavailability of various forms of aluminum and the transport and metabolism of aluminum within biological systems.' Many other medically related AMS studies have been reported and two of these will be mentioned here. The first is the development on the AMS facility at Oxford University of an ion source microprobe for mapping carbon-14 in biological specimens [18] and the second, already discussed in Chapter 6, is the detection of chlorine-36 in samples of structural material from Hiroshima to determine the neutron doses received by survivors during the atomic bombing of that city in 1945. The data of Straume *et al* (see references [5–8], Chapter 6) demonstrated that, at distances beyond 1 kilometre or so, the main radiation damage to people was through neutrons and not gamma rays as previously believed. The populations of both Hiroshima and Nagasaki who were exposed to the A-bomb radiation in 1945 are the best studied in the world in terms of radiation effects. After almost 50 years it is finally possible to estimate with some accuracy how great a radiation dose they, in fact, received.

The detection of tritium—a radioisotope of hydrogen with a half-life of 12 years—by AMS is being developed at Lawrence Livermore [19] to provide high-sensitivity tritium measurements in milligram biological samples. Labelling of such samples with both tritium and carbon-14 will permit the performance of unique double labelling experiments. Also the commercial availability of a large number of tritium-tagged compounds will allow experiments with chemicals that are not available in a radiocarbon-tagged form.

The detection of carbon-14 by AMS, however, continues to be the main biomedical application. A review paper by Vogel and Turteltaub [20] of the Lawrence Livermore National Laboratory presented at the Sixth International Conference on Accelerator Mass Spectrometry held in Australia in September–October 1993 mentions several other programmes designed to understand the kinetics and specific binding properties of carcinogens and toxins. These two authors note that, despite the expansion in biomedical applications of AMS, these studies remain a relatively small fraction of the world effort in AMS. They further point out that, although it appears that the biomedical community is slower to adopt AMS than the geo-chronometry community, the learning curves are similar. This can be seen by comparing the radiocarbon dating by AMS in 1982, five years after its introduction, to the

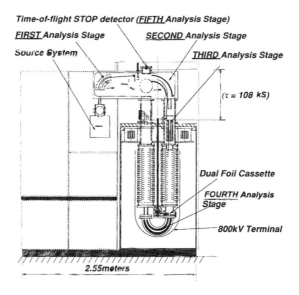

Time-of-flight STOP detector (*FIFTH* Analysis Stage)

*FIRST* Analysis Stage          *SECOND* Analysis Stage

Source System          *THIRD* Analysis Stage

(τ = 108 kS)

Dual Foil Cassette

*FOURTH* Analysis Stage

800kV Terminal

2.55meters

*The most recent design for a tandem AMS system 'Attomole 2100' for biomedical and environmental applications. It has a terminal potential of 800 kilovolts and has the compact dimensions of 2.5 metres (height), 2.5 metres (length) and 1.25 metres (width). Molecules are destroyed in the terminal at this relatively low terminal voltage by virtue of the use of foil strippers. Courtesy of K H Purser, Southern Cross Corp., Peabody, MA.*

biomedical use of AMS in 1993, five years after the initial efforts in 1988. They write: "The obstacles to acceptance are nearly identical in both instances: accelerator facilities for biomedical AMS are few, and access seems limited to in-house research.' They stress the importance of increasing the number of AMS facilities that devote time to biomedical research, and especially broadening that research to uses of radioactive tracers in addition to carbon-14 such as tritium, aluminium-26 and calcium-41.

Present AMS facilities, even those specifically designed for radiocarbon dating, are either larger than required for biomedical work or are tailored to make more accurate measurements than needed in the field. However, accelerators have now been designed specifically for biomedical research [21, 22].

In summary, the use of AMS to detect radioactive tracers in biomedical research, firstly, greatly reduces the radioactive tracer dose that needs to be injected into the subject as well as the blood, urine or other samples that need to be collected for study, secondly, the radioactive waste disposal problem is minimal or non-existent and, thirdly, the tracer dose to human subjects is so low it can readily meet government regulations. One can confidently predict in the not too distant future burgeoning applications of AMS in biomedical research. The step after that will be the use of AMS in medical diagnostics, most probably in the field of chemotherapy. It may be possible, for example, with AMS to predict the declining efficacy of a particular chemotherapeutic drug well before its deleterious side effects can occur.

The interest in measuring chlorine-36 by AMS originally stemmed from the belief that it would provide an important way to measure the age of very old ground water. This was mentioned in Chapter 3 where the workshop on dating ground water was cited (reference [4], Chapter 3). The organizers of the workshop were concerned with the question that, to this day, plagues many members of the scientific community, how should one deal with the immense quantities of high-level, long-lived nuclear waste that has accumulated since the nuclear age began in World War II? As noted in the introduction to the workshop report, 'The dating of ground water in the vicinity of nuclear repositories is a critical aspect of the safety evaluation of these repositories. The age of the water gives an indication of past rates of water movement which in turn will help evaluate the likelihood of future migration of radionuclides from a given repository.' It should be noted that the age of ground water is defined as the length of time the water has been isolated from the atmosphere.

The advantages of chlorine-36 for ground water dating are its suitable half-life of 0.301 Ma (million years), its simple geochemistry, its conservation in ground water (once incorporated into a water system the chlorides stay there and, in general, no chlorides enter ground water aquifers from the environment through which the water is transported) and a general absence of subsurface sources at levels comparable to the atmospheric input. When it was established that chlorine-36 could be measured by AMS with considerable precision and, more importantly, sensitivity, the search began for a way of conclusively demonstrating that it could be employed to measure the age of ground water on as much as a million year time scale. What was needed to test the efficacy of using the measurement of chlorine-36 to date ground water was to find an aquifer that was sufficiently well understood that its flow rate could be estimated from hydrodynamic simulations. The Great Artesian Basin in Australia seemed ideal. It is one of the largest artesian aquifer

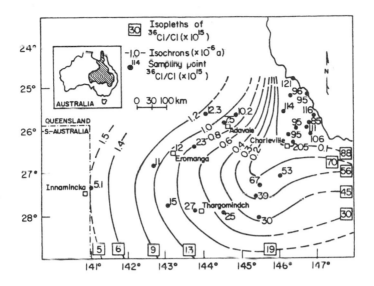

*Map of the Great Artesian Basin area of Australia. The water sampling points are shown as full circles. The full curves are isopleths of the $^{36}Cl/Cl$ ratio converted to isochrons based on the half-life of chlorine-36. The inset map of Australia shows where the basin is located (crosshatched).*

systems in the world, occupying an area of almost 2 million square kilometres, about one-fifth of Australia. The hydrology of this basin is relatively simple and so the computed hydrological ages are concomitantly credible.

A map of the Great Artesian Basin shows its place in Australia and the location of the water sampling points. The solid curves are isopleths of $^{36}Cl/Cl$ ratios that are converted to isochrons labelled in units of 1 Ma (million years) based on the half-life of chlorine-36. An isopleth is a line on a map on which some quantity (in this case the $^{36}Cl/Cl$ ratio) is constant. A photograph taken at one of the drill sites shows a rusting steam engine that was hauled in from the coast over 1000 kilometres away at the turn of the century to drill a well almost 1 mile deep. (See the colour section.) It was abandoned when the well was completed. Another photograph shows a water sampling operation in progress. The water temperature was 33 °C.

From measurements of the chlorine radioisotope to stable isotope ratio ($^{36}Cl/Cl$) at various points in the aquifer compared with the ratio at the recharge location and the amount of chlorine per litre of water at the same points (from these two quantities

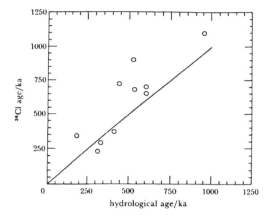

*The chlorine-36 age in thousands of years (ka) versus the hydrological age in ka. A one to one correspondence between the two would fit the 45° line shown. A better fit is to a 51° line.*

the amount of chlorine-36 per litre can be calculated) one can obtain the chlorine-36 age of the water in, for example, kilo (thousands of) years (ka). For example at a sampling location where the number of chlorine-36 atoms per litre of water is half the value in the recharge location the water is a third of a million years old. This chlorine-36 age can then be plotted against the hydrological age in ka and the results are shown in the plot. A one to one correspondence between the chlorine-36 age and the hydrological age would require the data to fit a line of 45° slope as shown. Although a better fit is to a line of slope about 51°, this is reasonably convincing evidence that the chlorine-36 atoms per litre in water compared with the value at recharge is a good measure of the length of time the water has been isolated from the atmosphere on a time scale of a million years [23–25].

This method for dating old ground water has, however, seen few applications. The technique of locating the bomb pulse of chlorine-36 in ground water to yield a flow rate, discussed in Chapter 9, turns out to be much more useful. If one knows the recharge area of an aquifer and can locate, along the flow direction of the aquifer, the place where the chlorine-36 concentration reaches a high maximum value, one has found the location of the pulse of chlorine-36 that was injected into the aquifer recharge area between 1951 and 1958. Knowing, say 40 years later, that the bomb pulse is a known distance further along the aquifer flow immediately measures the flow rate of the aquifer.

There are many applications of AMS measurements of chlorine-

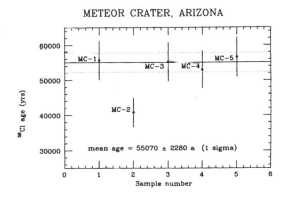

METEOR CRATER, ARIZONA

*The build-up of chlorine-36 on the surface of five boulders thrown up by the meteorite impact and resting on the rim of the crater was measured by AMS and established that the impact occurred 55 000 years ago.*

36 in geology. One of these is a determination of the exposure ages of rock surfaces. The technique is based on the fact that, when a rock surface is exposed to the atmosphere, chlorine-36 (produced by the spallation of K and Ca and thermal neutron capture on chlorine-35) accumulates in the rocks owing to cosmic rays bombarding the surface of the earth. The build-up is a function of time, location and composition of the rock. The method can be used to directly date glaciations and meteorite impact times. In the case of glaciations the rock surfaces scoured by glaciers become exposed to cosmic rays at the earth's surface when the glacier retreats. The amount of chlorine-36 build-up in the rock signals the time of the glacial retreat. In the case of meteorite impact, rocks deeply buried below the surface and thus shielded from cosmic rays may be raised to the crater rim and become exposed to cosmic rays. The build-up of chlorine-36 then begins.

The Meteor Crater in Arizona (formerly called Canyon Diablo Crater), located on the Colorado Plateau, 55 kilometres east of Flagstaff, Arizona, resulted from the impact of an iron–nickel meteorite. It is approximately 1.1 kilometres across and 200 metres deep. Samples were obtained from the surfaces of five boulders on the crater's rim and the chlorine-36 content of material at and near the rock surfaces was measured at Rochester [26]. The method of chlorine-36 build-up enabled a determination of the time that the rock surfaces were first exposed. Four of the boulders yielded an average age of $55.1 \pm 2.2$ ka (thousand years). The fifth boulder was appreciably younger giving an age of about 40 ka. This younger age could be due to the fact that the boulder rolled over or shattered, exposing a fresh surface.

Turning now from the exposure ages of rocks to biblical ages we consider the fascinating saga of the Dead Sea scrolls [27]. They are an exemplar of the advantages of AMS for radiocarbon dating. Samples of Dead Sea scroll material comprising parchment, papyrus and linen weighing a few tens of milligrams were all that could be made available for carbon dating. Samples of this size were just too small to be dated by the Libby decay counting method. As a result, although they were first discovered in the mid twentieth century it was almost 50 years later that the were reliably dated by AMS by Bonani *et al* at the AMS laboratory in Zurich, Switzerland [28, 29] and Jull *et al* at the University of Arizona AMS laboratory in Tucson, Arizona [30].

In 1947, young Bedouin shepherds, searching for a stray goat in the Judean Desert, stumbled upon a long unvisited cave near the northwest shore of the Dead Sea. In it they found jars filled with ancient scrolls. Eventually thousands of scroll fragments were discovered in eleven caves in cliffs about a mile from the shore at the north end of the Dead Sea near an old ruin named Khirbet Qumran some 20 kilometres west of Jerusalem. According to Jull *et al* [30]: 'The Qumran scrolls are generally considered to have been hidden by the Qumran Community, identified by most scholars as the Essenes. The documents are usually regarded to have been copied between the mid-third century BC and AD 68, when the Qumran settlement was destroyed by the Romans.'

Ever since their discovery, scholars have been engaged in the daunting task of assembling the fragments in what comprises a giant jigsaw puzzle. They are written in three different languages, Hebrew, Aramaic and Greek, and constitute the greatest manuscript find of the twentieth century. According to Bonani *et al* [29]: 'The name Dead Sea Scrolls refers to some 1200 manuscripts found in caves in the hills on the western shore of the Dead Sea during the last 45 years. They range in size from small fragments to complete books from the holy scriptures (the Old Testament). The manuscripts also include uncanonized sectarian books, letters and commercial documents, written on papyrus and parchment. In only a few cases was direct information on the date of writing found in the scrolls. In all other cases the dating is based on indirect archaeological and paleographical evidence.'

Bonani noted that although the radiocarbon dating method

was developed by Libby at approximately the same time as the scrolls were discovered, too much material (several grams) would be required to directly date the scrolls themselves. Related material such as linen wrapping of a scroll and charred date palm logs excavated at the Qumran site were dated by the Libby decay counting method indicating that the scrolls might be at least 2000 years old. Dating of material from the scrolls themselves became feasible only after the development of AMS which permits the dating of material containing a milligram or so of carbon. Palaeography, the study of ancient writings, considered to be an accurate method of dating, suggested that the scrolls dated between the third century BC and AD 68 when the site was abandoned. However, in the decades following the initial discovery of the scrolls a number of scholars began to challenge the palaeographic datings. For this reason Bonani *et al* felt it would be useful to check on the palaeographically determined ages by an independent method, i.e. radiocarbon dating by AMS.

Employing the AMS facility in Zurich, they dated 14 fragments of scroll material, both papyrus and parchment [29]. The uncertainties in the radiocarbon ages achieved by the Zurich group on these 14 samples were ±38 years on average for material some 2000 years old, an impressive accomplishment indeed. Eight of these were parchment samples taken from the Qumran caves and the AMS results yielded dates from 300 BC to AD 61. Four papyrus samples from sites other than the Qumran caves but in the vicinity came from scrolls that bore actual dates. The AMS dates agreed remarkably well with the actual dates within $1\sigma$ limits (see below). Palaeographic dating was vindicated but additional support was to be forthcoming.

This support came from scientists at the National Science Foundation Arizona Accelerator Mass Spectrometry Facility at the University of Arizona in Tucson, Arizona. On 21 and 22 March 1994 scroll samples were selected by museum staff at the Rockefeller and Israel Museums in Jerusalem in the presence of the Arizona scientists, Jull *et al* [30]. Three additional samples were taken later from the Israel Museum and sent to Tucson. All samples were taken from ragged edges of top or bottom margins of the scrolls. No samples were taken that would have caused significant damage to the scrolls themselves. The sizes of the samples ranged from about 5 to about 57 milligrams, averaging about 18 milligrams per sample. Three of the samples were date-bearing documents whose identity and ages were unknown to the Arizona researchers. One of the samples had been previously dated by Zurich and its identity was also unknown to Arizona. There were two linen samples, one from Qumran cave 4, weighing 31 milligrams and the other, purportedly, from cave 2, weighing 22 milligrams.

The sample dates were quoted in terms of $1\sigma$ and $2\sigma$ limits. For example, a sample with a listed $1\sigma$ age between AD 14 and 115 means there is a 68% probability that its age lies between these limits. In this particular case the $2\sigma$ age is 43 BC to AD 141 and this means there is a 95% probability that its age lies between these limits. For the three date-bearing samples two of the Arizona dates agreed within the $1\sigma$ limit and one was outside the $2\sigma$ limit by only 10 years. The Arizona result for the Zurich sample agreed with the Zurich value almost exactly. The authors conclude: 'Measurements on samples of known age are in good agreement with those known ages. Ages determined from $^{14}$C measurements on the remainder of the Dead Sea Scroll samples are in reasonable agreement with paleographic estimates of such ages, in cases where those estimates are available.'

The average uncertainty achieved by the Arizona group in the some 20 ages they reported for Dead Sea scrolls and related materials was $\pm39$ years in samples that were some 2000 years old. In this respect they emulated their Zurich colleagues.

Finally the results [30] for the two linen samples are of interest. One of these weighing 31 milligrams was a sample of cloth from Qumran cave 4, to which a leather thong was attached of the kind used to fasten the scrolls at Qumran. Its $1\sigma$ age range was 160–41 BC. This fell solidly within the dating period established for the scrolls from cave 4 both by palaeography and carbon-14 dating. The other weighing 22 milligrams, a linen fragment with silk embroidery, dated to the twelfth to thirteenth centuries AD. This sample was bought from antiquity dealers who represented it as material from Qumran cave 2. Textile experts agree that it is unlikely to be Qumran cave material but rather from some medieval source.

My particular interest in the linen dates arises from the suggestion of Garza-Valdes (see reference [12], Chapter 10) that bacterial infestation of ancient cloth can produce a bioplastic coating. Such a coating can incorporate more recent carbon-14 into the material. This coating is not removed by the standard cleaning process used by carbon dating laboratories. It thus can cause the cloth radiocarbon dates to appear to be substantially younger than the date the flax (in the case of linen cloth) was harvested. This is patently not the case for the Qumran cave 4 cloth sample. Although, in principle, it could explain the date for the second sample the questionable relationship of that sample to the scrolls makes this possibility extremely unlikely. It should be noted, however, that no microscopic examination was made of either of the two linen samples so there is no evidence one way or the other of the existence of any bioplastic coating.

# Chapter 12

# Some Thoughts on Future Developments of AMS

Since the genesis of AMS in the late 1970s it has been applied to numerous areas of science. Many of these have been described in this book and in review articles listed in references in previous chapters. They include archaeology, atmospheric science, atomic, nuclear and particle physics, biomedicine, condensed matter, geoscience, glaciology, hydrology and oceanography. To predict the future of a technique so replete in examples of its present applicability is somewhat daunting. There are, however, several future growth areas that seem obvious, some of which are currently in their infancy. These are, not necessarily in order of importance, biomedicine, condensed matter, $CO_2$ gas sources for radiocarbon detection, the use of cesium ion source microprobes, archaeology, the detection of actinides and the use of neutral injection for tandem electrostatic accelerator systems (NI AMS). They are discussed in more detail below.

### Applications of AMS to Biomedicine

There are a number of radionuclides that could make useful tracers in medical research and which are readily detectable at extremely low concentrations by AMS. They include tritium ($^3$H), carbon-14, aluminium-26, chlorine-36, calcium-41 and iodine-129. Another radionuclide that may be of interest is selenium-79 but no effort has yet been devoted to its detection by AMS.

Of these carbon-14 is the most universally used biochemical tracer and is the one for which the capability of detection by AMS is most developed. It is also the one that is most easily, accurately

196

and sensitively measured by tandem accelerators having terminal potentials of the order of 1 MV. Some examples of the detection of carbon-14 as well as calcium-41, aluminium-26 and tritium by AMS in biomedicine are given in the previous chapter [1].

The day that AMS will be used by the medical profession for early diagnosis of certain diseases cannot be far off. Its present somewhat limited application to the study and understanding of important human illnesses is bound to greatly increase.

If the biomedical research using AMS is to be more widely performed smaller and cheaper AMS facilities will be required. The general parameters of such a facility for the detection of carbon-14 in biological samples have been mentioned [2] in the previous chapter (see the figure on page 187).

## Applications of AMS in the field of condensed matter

Such applications are still in their early stages and so have not been discussed in the present book. An early example of the power and sensitivity of AMS to detect a wide range of trace impurities in semiconductors was described by Anthony and Donahue [3]. About the same time, and independently, the depth profiling of chlorine and nitrogen in ultra pure silicon wafers was demonstrated using a combination of neutron activation and AMS [4] and more recently the diffusion of chlorine in silicon wafers was studied [5]. The application of AMS to silver halide imaging has been explored [6] although the rapid development of solid state technology for photographic imaging is reducing the importance of such research.

Wherever trace elements play a role in the performance of solid state devices or materials their quantitative measurement by AMS should become increasingly important in the future.

## Carbon dioxide gas sources for radiocarbon detection

For most measurements of carbon-14 in carbonaceous samples by AMS the carbon in the sample is extracted as $CO_2$ and then converted to solid graphite for use in the negative ion sputter sources used with tandem accelerator AMS systems. If this graphitization step could be circumvented and the $CO_2$ be used directly in the ion source several advantages would accrue. These include [7, 8] a saving in target preparation time, a higher efficiency, the ability to maintain the same carbon current for a wide range of sample sizes and direct monitoring of background contributions which allow these to be estimated to a high degree of certainty. In addition it would eliminate the disadvantages that the process of forming graphite can introduce isotopic fractionation (changes in the $^{14}C/^{13}C/^{12}C$ ratios), that it can introduce carbon-14 contamination and that the physical properties of the graphite formed from different samples is rarely identical. The disadvantages of a gas source are that the sputtering of material initially in a gas phase is less well understood than sputtering from solid targets and, since the $CO_2$ gas is not physically confined, there is the greater possibility of cross contamination from one sample to another (the so-called memory effect).

Important contributions to the development of carbon dioxide gas sources for tandem accelerators have been made by the AMS group at Oxford University (see [8] and references therein). In this source the $CO_2$ is adsorbed on a solid titanium surface and $C^-$ beams are produced by sputtering the titanium surface with a $Cs^+$ beam. Different titanium targets are used for each gas sample to reduce the memory effect. Solid graphite targets can also be used in the case of the hybrid (gaseous and solid) source [8].

Another approach is to inject gas samples directly into the plasma of a compact microwave negative ion source (CMS). This source was developed [9, 10] at the Chalk River Laboratories of Atomic Energy of Canada, Ltd for the production of heavy ion beams for the Tandem Accelerator Super Conducting Cyclotron (TASCC) at that laboratory. When that accelerator project was terminated in 1997 the source was given to the AMS laboratory at the University of Toronto. It is currently on loan by Toronto to the AMS laboratory at Woods Hole, Massachusetts. If this source proves to perform as required it may well become the gas negative ion source of choice for tandem AMS systems measuring carbon-14 to carbon ratios.

### Cesium ion source microprobes

Cesium beams in AMS negative ion sources of sufficient intensities and with diameters approaching 50 micrometres or less are being developed for use in scanning polished surfaces of geological and other samples to determine the distribution of important elements in tiny inclusions in those samples. Very preliminary work in this area was carried out in the early 1980s at the University of

Rochester [11] but the technique is now being refined on the AMS facility at the University of Toronto. Its most important application will occur when combined with neutral injection AMS (NI AMS) as discussed below.

**Archaeology**

To the general public the application of AMS to the carbon dating of ancient organic archaeological artefacts holds the greatest fascination. One is still not sure, for example, how, when and by whom the Americas were first populated. AMS has provided some answers and it is certain that it will continue to play an important role in attempting to solve this and many other archaeological puzzles.

**The detection of actinides**

The application of AMS to the ultrasensitive detection of the actinides is also at an early stage. AMS laboratories in Canada [12, 13] and Australia [14, 15] have done pioneering work in this area. As an example, the detection of uranium-236 is of interest because of its production in the enriched fuels employed in certain types of reactor cores, for example those in nuclear powered ships. Heavy element detection by AMS has substantial growth potential which may well be aided by the development of NI AMS discussed below.

**Neutral injection for AMS (NI AMS)**

(a) *Overview*
In 1988 Kutschera [16] noted that there are more than 140 radioisotopes with half-lives longer than 1 year. Only a handful of these have been measured by AMS and, at the time that article was written, with the exception of iodine-129, that handful included no radioisotopes heavier than mass about 50. Not much has changed since then. If, in addition to these radioisotopes, one includes stable isotopes, the number of isotopes that could be, but have not yet been, detected by AMS, is very large. Part of the reason for this is that the electron affinity (the binding energy of an electron to a neutral atom) for the element may be too low to make it possible to efficiently create negative ion beams of the element for injection into tandem accelerator AMS systems. In such cases

neutral injection provides a solution. It is also the solution when the element that is wanted is embedded in an insulating matrix as will be discussed below. Neutral injection for nuclear physics applications was first suggested by Van de Graaff [17] and its application to AMS was first proposed by Purser *et al* [18] in 1979 and by Litherland and Kilius in 1997 [19]. It may well constitute the next revolution in AMS.

(b) *Outline of an NI AMS system*
A suitable negative ion beam such as $I^-$ would be used in an ion source to directly produce singly charged positive ions of the element to be measured. These ions would be accelerated to about 100 keV and then focused to have a waist at the stripping canal in the centre of the tandem accelerator. They would then immediately be converted to a high-quality neutral beam with high efficiency (>70%) by resonant electron transfer in a charge exchange cell containing a suitable gas. This beam of neutrals would be injected into the tandem where they would drift without further acceleration or deflection to the terminal stripping canal. In this canal they would be converted again to singly charged positive ions and accelerated down the second half of the tandem to ground. There they can be charge changed efficiently [20–22] to 3+ in a gas cell or foil to completely eliminate molecules and then further analysed by appropriate combinations of magnetic and electrostatic analysers.

(c) *Discussion of the advantages of neutral injection for AMS (NI AMS)*
When the electron affinity of a nuclide is low or when the wanted nuclide is embedded in an insulator the initial production of singly charged positive ions directly is highly advantageous in an ion source or by bombardment by a negative ion beam such as $I^-$. Singly charged positive ions are readily produced for most atoms and molecules, unlike the situation for negative ions.

When the wanted atoms are embedded in an insulator, for example metals in a silicate rock matrix, the use of singly charged positive cesium ions to bombard the insulator to produce the required singly charged negative ions produces a build-up of charge. This causes voltage instabilities that invalidate the measurements. When negative ions are employed to bombard the insulator to produce the required singly charged positive ions there is little or no build-up of charge. This is probably due to the mobility of the electrons created by the negative ion bombardment and their being driven off by the electric field of the negative ions.

Neutral injection of atoms and molecules of the same mass to the terminal of a tandem followed by a charge change to 1+ in the terminal and then followed by a charge change to 3+ at ground (needed for molecular break-up of the molecular isobars) has an advantage over standard AMS (negative ions into the tandem followed by charge change to 3+ in the terminal). It affords a better mass energy product separation between the required atom and the molecular fragments from the molecular isobar.

### (d) *General applications of NI AMS*

The *platinum group elements* (PGEs), ruthenium, rhodium and palladium ($Z = 44, 45, 46$) and osmium, iridium and platinum ($Z = 76, 77, 78$), are all transition elements. They all have at least one abundant stable isotope with no stable isobar. They have electron affinities that range from 0.6 to 2.1 eV and ionization potentials that range from 7.4 to 9.1 eV. Their detection at very low levels in insulating matrices such as rocks is of interest. Except for iridium they all have at least one radioisotope with a half-life exceeding 1 year but all of these have a stable isobar. Platinum and iridium were originally measured in clay from the Cretaceous–Tertiary boundary and in untreated rock and mineral material (for early measurements of the PGEs by AMS see [11, 23, 24]).

Subsequent extensions of and improvements to AMS measurements of the PGEs and other metals have been carried out at IsoTrace (the University of Toronto's AMS laboratory). These involved the preparation of polished mounts of conducting targets of coarse-grained samples (grain sizes of 1 millimetres and up) for *in situ* analysis. Polished surfaces of sulphides, oxides, graphite, copper metal and meteoritic Ni–Fe alloys were optically scanned prior to analysis, in order to select the best target grains and to avoid unwanted impurities. Although a compromise solution involving the broad $Cs^+$ primary beams of sputter ion sources meant primarily for radionuclide measurements, this approach has opened avenues of research into a wide variety of ore minerals and other natural conductors [25–38].

Due to the siderophilic (found in the metal phase) and chalcophilic (found in the sulphide phase) nature of the PGEs they preferentially enter the sulphide phase of a molten rock and thus leave the siliceous phase depleted. Thus, in layered intrusions, the PGEs are found primarily in sulphide concentrations with very low levels of PGEs in the silicates. The PGEs should be enriched in the sulphide layer and have low levels in the silicates above these layers. The upper silicate layers will be depleted in PGEs since they were scavenged by the sulphide droplets as they settled through it. The lower silicate layer, not stripped of PGEs in this manner, will reflect the PGE levels of the magma that has not equilibrated with the sulphides. The studies of the PGEs in layered intrusions will provide information of interest to economic geologists [39, 40] and on certain fundamental geological processes.

It has been pointed out [41] that the problem is to analyse the silicates accurately. Bulk chemical analysis of mineral separates (by powdering and mixing with a conductor such as a very pure silver powder) is inadequate since minute inclusions of sulphides will invalidate the result. The only method suitable for the PGEs in silicates is *in situ* measurements by AMS. AMS with detection limits in the range of 0.1 parts per billion (ppb) or less offers adequate sensitivity. However, such measurements in insulators such as silicates are not possible by the usual AMS technique of producing negative ions of the PGEs by bombardment of the silicates with beams of $Cs^+$ ions. The build-up of positive charge that would occur in the silicate insulator would cause voltage instabilities and these instabilities preclude the possibility of making meaningful measurements. The solution is to employ NI AMS as described in (b) above.

The *lanthanides* (sometimes called rare earth elements) are elements with $Z = 58$ to 71. Sometimes lanthanum with $Z = 57$ is included. They all have very low (less than 0.5 eV) or negative electron affinities and ionization potentials that range from 5.4 to 6.3 eV. They all have at least one stable isotope with no stable isobar. Many have practical applications. Except for praseodymium, erbium and ytterbium, they all have at least one radioisotope with a half-life exceeding one year. Again most of these have a stable isobar. For example samarium-146 has a half-life of $10^8$ years. It is present in meteorites. There is one stable isobar neodymium-146, but there are ways of substantially reducing neodymium-146 with respect to samarium-146.

The transition elements other than the PGE discussed above include elements with $Z = 21$–30, $Z = 39$–48 and $Z = 72$–80. Sometimes lanthanum with $Z = 57$ is included. Except for iridium, platinum and gold they all have electron affinities of

1.3 eV or less. Several have negative electron affinities, probably including technetium. Technetium-99 with a half-life of $2.1 \times 10^5$ years is of particular interest. It is an abundant light fission product and has no crustal or cosmogenic origin. It could serve as a sensitive tracer of leakage from storage facilities of waste from nuclear reactors. It has a stable isobar, ruthenium-99, but ways can probably be devised to separate the two isobars.

The *noble gases* include neon, argon, krypton, xenon and radon ($Z = 10, 18, 36, 54, 86$). None of them form negative ions. Their ionization potentials range from 11 to 22 eV. Except for radon they all have at least one stable isotope that has no stable isobar. There are no radioisotopes of neon with half-lives of more than one year. Only argon-39 and argon-42 are long-lived radioisotopes of argon, with half-lives of 269 and 33 years respectively. They both have stable isobars potassium-39 and calcium 42 respectively. In the case of krypton there are two long-lived radioisotopes namely krypton-81 and krypton-85 with half-lives of $2 \times 10^5$ years and 10.7 years respectively. They have stable isobars bromine-81 and rubidium-85 respectively. There are nine stable isotopes of xenon. Only xenon-131 has no stable isobar. There is no radioisotope of xenon with a half-life of more than one year. In the case of radon there is no stable isotope and no radioisotope with a half-life of greater than one year. There are many potential examples of applications of AMS to noble gases including the measurement of krypton-85 with a half-life of 10.4 years by AMS to detect clandestine nuclear weapons tests [42, 43].

*Cesium* has only one stable isotope, cesium-133 and there is no stable isobar. There are three long-lived radioisotopes, cesium-134, cesium 135 and cesium-137 with half-lives of 2, $3 \times 10^6$ and 30 years respectively. Each has a stable isobar of barium and cesium-134 has a stable isobar of xenon in addition. It is easier to make a singly charged positive cesium ion than a singly charged negative ion. Cesium-135 and cesium-137 are products of the fission of uranium-235 at an abundance of about 6%. The $^{135}Cs/^{137}Cs$ ratio can be used as an ocean tracer.

The *actinides* comprise all the elements with $Z$ greater than or equal to 89. The isotopes of all the actinides are radioactive. Their electron affinities are low, in general probably less than about 0.5 eV. They all have ionization potentials of 6 to 7 eV. One actinide of particular interest for detection by AMS is uranium-236 with a half-life of $2.34 \times 10^7$ years. Measurement of the $^{236}U/^{238}U$ ratio is particularly favourable for NI AMS as there is no mass 236 isobar formed naturally and the rate due to the environmental neutron flux, from the spontaneous fission of uranium-238 and the ($\alpha$, n) reactions of the alpha particles from uranium and thorium, is about $10^{-14}$ or about 10 ppq (parts per quadrillion, or 1 part in $10^{15}$). From present experience this ratio should be obtainable by NI AMS using a pair of magnets and their associated electric analysers. This ratio would represent a frontier in NI AMS and is well beyond what is thought to be necessary for the study of PGE in geological samples.

In conclusion, this book on the development of AMS and many of its applications is far from definitive. Nor is it a technical account of the subject. Such an account has been written by Tuniz *et al* (see reference [3], Chapter 1). The present book is designed to introduce the technique of AMS to potential users and to the general public. The applications included by no means constitute a complete set of AMS research areas nor a complete account of a particular application. They are, however, ones of wide appeal and ones of particular interest to the author. Accelerator mass spectrometry, in a real sense, is still in a developmental stage, principally because of its extraordinary sensitivity, the variety of its potential applications and because it is a technique that is still not familiar to many members of the scientific community. Undreamed of opportunities and applications undoubtedly still lie ahead.

# References

## References to Chapter 1

[1]   Gove H E 1996 *Relic, Icon or Hoax? Carbon Dating the Turin Shroud* (Bristol: Institute of Physics Publishing)
[2]   For a brief history of the invention of carbon-14 dating by W F Libby see Arnold J R 1992 The early years with Libby at Chicago: a retrospective *Radiocarbon After Four Decades* (New York: Springer) pp 3–10
[3]   Tuniz C, Bird J R, Fink D and Herzog G F 1998 *Accelerator Mass Spectrometry 'Ultrasensitive Analysis for Global Science'* (Boca Raton, FL: CRC Press)

## References to Chapter 2

[1]   Libby W F 1946 Atmospheric helium three and radiocarbon from cosmic radiation *Phys. Rev.* **69** 671–2
[2]   Oeschger H, Houtermans J, Loosli H and Whalen M 1970 The constancy of cosmic radiation from isotope studies in meteorites and on earth *Proc. 12th Nobel Symp., Radiocarbon Variations and Absolute Chronology, 1969* ed I U Olsen (New York: Wiley) pp 471–96
[3]   Purser K H 1977 Ultra-sensitive spectrometer for making mass and elemental analyses *US Patent 4037 100* (filed 1 March 1976)
[4]   Muller R A 1977 Radioisotope dating with a cyclotron, *Science* **196** 489–94
[5]   Alvarez L W and Cornog R 1939 Hu³ in Helium *Phys. Rev.* **56** 379
[6]   Muller R A, Alvarez L W, Holley W R and Stephenson E J 1977 Quarks with unit charge: a search for anomalous hydrogen *Science* **196** 521–3
[7]   Stephenson E J, Alvarez L W, Clark D J, Gough R A, Holley W R, Jain A and Muller R A 1977 *Bull. Am. Phys. Soc.* **22** 579
[8]   Schwarzschild A Z, Thieberger P and Cumming J B 1977 *Bull. Am. Phys. Soc.* **22** 94
[9]   Purser K H, Liebert R B, Litherland A E, Beukens R P, Gove H E, Bennett C L, Clover M R and Sondheim W E 1977 An attempt to detect stable N⁻ ions from a sputter ion source and some implications of the results for the design of tandems for ultra-sensitive carbon analysis *Rev. Phys. Appl.* **12** 1487–92
[10]  Crandell D R, Mullineaux D R and Rubin M 1975 Mount St Helens volcano: recent and future behavior *Science* **187** 438–41
[11]  Anbar M 1978 The limitations of mass spectrometric radiocarbon dating using CN⁻ ions *Proc. First Conf. on Radiocarbon Dating with Accelerators*

*(University of Rochester)* ed H E Gove pp 152–5 and references therein (unpublished)

[12]   Nelson D E, Korteling R G and Stott W R 1977 Carbon-14: direct detection at natural concentrations *Science* **198** 507–8

[13]   Bennett C L, Beukens R P, Clover M R, Gove H E, Liebert R B, Litherland A E, Purser K H and Sondheim W E 1977 Radiocarbon dating using electrostatic accelerators: negative ions provide the key *Science* **198** 508–10

[14]   Bennett C L, Beukens R P, Clover M R, Elmore D, Gove H E, Kilius L, Litherland A E and Purser K H 1978 Radiocarbon dating with electrostatic accelerators: dating of milligram samples *Science* **201** 345–7

## References to Chapter 3

[1]    Gove H E (ed) 1978 *Proc. First Conf. on Radiocarbon Dating with Accelerators (University of Rochester)*

[2]    Gove H E, Fulton B R, Elmore D, Litherland A E, Beukens R P, Purser K H and Naylor H 1979 Radioisotope detection with tandem electrostatic accelerators *IEEE Trans. Nucl. Sci.* **26** 1414–21

[3]    Purser K H, Litherland A E and Gove H E 1979 Ultra-sensitive particle identification systems based upon electrostatic accelerators *Nucl. Instrum. Methods* **162** 637–56

[4]    Davis S N 1978 *Workshop on Dating Old Groundwater* Y/OWI/SUB-78/55412 (Tucson, AZ: University of Arizona)

[5]    Kutschera W (ed) 1981 *Symp. on Accelerator Mass Spectrometry (Argonne, IL)* (unpublished)

[6]    Woelfli W, Polach H A and Anderson H H (ed) 1984 *Proc. Third Int. Conf. on Accelerator Mass Spectrometry (Nucl. Instrum. Methods* B **5** 91–488)

[7]    Gove H E, Litherland A E and Elmore D (ed) 1987 *Proc. Fourth Int. Conf. on Accelerator Mass Spectrometry (Nucl. Instrum. Methods* B **29** 1–445)

[8]    Yiou F and Raisbeck G M (ed) 1990 *Proc. Fifth Int. Conf. on Accelerator Mass Spectrometry (Nucl. Instrum. Methods* B **52** 211–630)

[9]    Fifield L K, Fink D, Sie S H and Tuniz C (ed) 1994 *Proc. Sixth Int. Conf. on Accelerator Mass Spectrometry (Nucl. Instrum. Methods* B **92** 1–524)

[10]   Jull A J T, Beck J W and Burr G S (ed) 1996 *Proc. Seventh Int. Conf. on Accelerator Mass Spectrometry (Nucl. Instrum. Methods* B **123** 1–594)

[11]   Elmore D, Gove H E, Beukens R P, Litherland A E, Purser K H and Rubin M 1978 A method for dating the Shroud of Turin, *La Sindone e la Scienza, Bilanci e Programmi: Atti del II Congresso Internazionale di Sindonologia, 1978* ed P Coero-Borga, Centro Internazionale (Turin: Edizioni Paoline) pp 428–31

[12]   Gove H E, Elmore D, Ferraro R, Beukens R P, Chang K H, Kilius L R, Lee H W, Litherland A E and Purser K H 1980 Radioisotope detection with tandem electrostatic accelerators *Nucl. Instrum. Methods* **168** 425–33

[13]   Elmore D, Fulton B R, Clover M R, Marsden J R, Gove H E, Naylor H, Purser K H, Kilius L R, Beukens R P and Litherland A E 1979 Analysis of $^{36}$Cl in environmental water samples using an electrostatic accelerator *Nature* **277** 22–5

[14]   Kilius L R, Beukens R P, Chang K H, Lee H W, Litherland A E, Elmore D, Ferraro R, Gove H E and Purser K H 1980 Measurement of $^{10}$Be/$^9$Be ratios using an electrostatic tandem accelerator *Nucl. Instrum. Methods* **171** 355–60

The first AMS measurements of $^{10}$Be/$^9$Be ratios were carried out at Grenoble, see Raisbeck G M, Yiou F, Fruneau M and Loiseaux J M 1978 Beryllium-10 mass spectrometry with a cyclotron *Science* **202** 215–17

[15] Nishiizumi K, Arnold J R, Elmore D, Ferraro, R D, Gove H E, Finkel R C, Beukens R P, Chang K H and Kilius L R 1979 Measurements of $^{36}$Cl in Antarctic meteorites and Antarctic ice using a Van de Graaff accelerator *Earth Planetary Sci. Lett.* **45** 285–92

[16] Finkel R C, Nishiizumi K, Elmore D, Ferraro R D and Gove H E 1980 $^{36}$Cl in polar ice, rainwater and seawater *Geophys. Res. Lett.* **7** 983–6

[17] Currie L A, Klouda G A, Elmore D and Gove H E 1985 Radiocarbon dating of microgram samples: accelerator mass spectrometry and electromagnetic isotope separation *Nucl. Instrum. Methods* B **12** 396–401

[18] Stuiver M and Kra R (ed) 1980 Tenth International Radiocarbon Conference *Radiocarbon* **22** 133–1016

[19] Elmore D, Gove H E, Ferraro R, Kilius L R, Lee H W, Chang K H, Beukens R P, Litherland A E, Russo C J, Purser K H, Murrell M T and Finkel R C 1980 Determination of $^{129}$I using tandem accelerator mass spectrometry *Nature* **286** 138–40

[20] Litherland A E 1980 Ultrasensitive mass spectrometry with accelerators *Ann. Rev. Nucl. Part. Sci.* **30** 437–73

[21] Kilius L R 1980 Ultra sensitive mass spectrometry using a tandem accelerator *PhD Thesis* Department of Physics, University of Toronto

[22] Alvarez L W, Alvarez W, Asaro F and Michel H 1980 Extraterrestrial cause for the Cretaceous-Tertiary extinction *Science* **208** 1095–108

[23] Officer C and Page J 1996 *The Great Dinosaur Extinction Controversy* (Reading, MA: Addison-Wesley)

[24] Rucklidge J C, Evensen N M, Gorton M P, Beukens R P, Kilius L R, Lee H W, Litherland A E, Elmore D, Gove H E and Purser K H 1981 Rare isotope detection with tandem accelerators *Nucl. Instrum. Methods* **191** 1–9

[25] Gove H E 1985 Accelerator-based ultrasensitive mass spectrometry *Treatise on Heavy-Ion Science* ed D A Bromley (New York: Plenum) pp 430–63

[26] Elmore D, Anantaraman A, Fulbright H W, Gove H E, Hans H S, Nishiizumi K, Murrell M T and Honda M 1980 Half-life of $^{32}$Si from tandem-accelerator mass spectrometry *Phys. Rev. Lett.* **45** 589–92

[27] Kutschera W, Henning W, Paul M, Smither R K, Stephenson E J, Yntema J L, Alberger D E, Cumming J B and Harbottle G 1980 Measurement of the $^{32}$Si half-life via accelerator mass spectroscopy *Phys. Rev. Lett.* **45** 592–6

[28] Muller R A, Stephenson E J and Mast T S 1978 Radioisotope dating with an accelerator: a blind measurement *Science* **201** 347–8

[29] Stephenson E J, Mast T S and Muller R A 1979 Radiocarbon dating with a cyclotron *Nucl. Instrum. Methods* **158** 571–7

## References to Chapter 4

[1] Gove H E, Kuehner J A, Litherland A E, Almqvist E, Bromley D A, Ferguson A J, Rose P H, Bastide R P, Brooks N and Connor R J 1958 Neutron threshold measurements using the Chalk River tandem Van de Graaff accelerator *Phys. Rev. Lett.* **1** 251–3

[2] Cockcroft J D and Walton E T S 1930 Experiments with high velocity positive ions *Proc. R. Soc.* A **129** 477–89; 1932 Experiments with high velocity positive ions (i) further developments in the method of obtaining high velocity positive ions *Proc. R. Soc.* A **136** 619–30; 1932 Experiments

with high velocity positive ions (ii) the disintegration of elements by high velocity protons *Proc R. Soc.* A **137** 229–42

[3]   Van de Graaff R J 1931 A 1,500,000 volt electrostatic generator *Phys. Rev.* **38** 1919–20; 1935 Electrostatic generator *US Patent 1991 236* (filed 16 December 1931)

[4]   Litherland A E, Paul E B, Bartholomew G A and Gove H E 1956 Gamma rays from the proton bombardment of $^{24}$Mg *Phys. Rev.* **102** 208–22
      See also Morinaga H 1956 Interpretation of some of the excited states of 4n self-conjugate nuclei *Phys. Rev.* **101** 254–8

[5]   Bohr A and Mottelson B R 1953 Collective and individual-particle aspects of nuclear structure *Dan. Mat. Fys. Medd.* **27** (16) 1–174

[6]   Nilsson S G 1955 Binding states of individual nucleons in strongly deformed nuclei *Dan. Mat. Fys. Medd.* **29** (16) 1–69

[7]   Gove H E and Litherland A E 1957 Comparison of the mirror nuclei $^{25}$Mg and $^{25}$Al *Phys. Rev.* **107** 1458–9

[8]   Van de Graaff R J 1960 Tandem electrostatic accelerators *Nucl. Instrum. Methods* **8** 195–202

[9]   Bennett W H and Darby P F 1936 Negative atomic hydrogen ions *Phys. Rev.* **49** 97–9
      Darby P F and Bennett W H 1936 The observation of negative hydrogen ions *Phys. Rev.* **49** 422; 1936 Negative hydrogen and deuterium ions *Phys. Rev.* **49** 881–2

[10]  Bennett W H 1953 Nuclear reactions between heavy nuclei *Rev. Sci. Instrum.* **24** 915–6; 1940 High voltage vacuum tube *US Patent 2206 558* (filed 9 July 1937)

[11]  Alvarez L W 1951 Energy doubling in dc accelerators *Rev. Sci. Instrum.* **22** 705–6

[12]  Alvarez L W 1981 The early days of accelerator mass spectrometry *Proc. Symp. on Accelerator Mass Spectrometry, 1981 (Argonne National Laboratory)* ANL/PHY-81-1, pp 1–15

[13]  Bromley D A 1974 The development of electrostatic accelerators *Nucl. Instrum. Methods* **122** 1–34

[14]  Middleton R 1974 A survey of negative ion sources for tandem accelerators *Nucl. Instrum. Methods* **122** 35–43

[15]  Middleton R 1989 *A Negative-Ion Cookbook* (University of Pennsylvania), revised 1990

[16]  Moak C D, Banta H E, Thurston J N, Johnson J W and King R F 1959 Duo plasma ion source for use in accelerators *Rev. Sci. Instrum.* **30** 694–9

[17]  Middleton R and Adams C T 1974 A close to universal negative ion source *Nucl. Instrum. Methods* **118** 329–36

[18]  Middleton R 1983 A versatile high intensity negative ion source *Nucl. Instrum. Methods Phys. Res.* **214** 139–50

## References to Chapter 5

[1]   Purser K H 1978 Accelerators—the solution to direct $^{14}$C detection *Proc. First Conf. on Radiocarbon Dating with Accelerators (University of Rochester)* ed H E Gove pp 1–32 (unpublished)

[2]   Purser K H, Litherland A E and Gove H E 1979 Ultra-sensitive particle identification systems based upon electrostatic accelerators *Nucl. Instrum. Methods* **162** 637–56

[3]   Litherland A E 1980 Ultrasensitive mass spectrometry with accelerators *Ann. Rev. Nucl. Part. Sci.* **30** 437–73

[4] Litherland A E 1984 Accelerator mass spectrometry *Nucl. Instrum. Methods Phys. Res.* B **5** 100–8

[5] Gove H E 1985 Accelerator-based ultrasensitive mass spectrometry *Treatise on Heavy Ion Science* vol 7, ed D A Bromley (New York: Plenum) pp 431–63

[6] Elmore D and Phillips F M 1987 Accelerator mass spectrometry for measurement of long-lived radioisotopes *Science* **236** 543–50

[7] Gove H E 1992 The history of AMS, its advantages over decay counting: applications and prospects *Radiocarbon After Four Decades, An Interdisciplinary Perspective* ed R E Taylor, A Long and R S Kra (New York: Springer) pp 214–29

[8] See, for example, Evans R D 1955 *The Atomic Nucleus* (New York: McGraw-Hill)

[9] Chew S H, Greenway T J L and Allen K W 1984 Accelerator mass spectrometry for heavy isotopes at Oxford (OSIRIS) *Nucl. Instrum. Methods Phys. Res.* B **5** 179–84

[10] Kutschera W 1997 Conference summary: trends in AMS *Nucl. Instrum. Methods Phys. Res.* B **123** 594–8

## References to Chapter 6

[1] Hersey J 1946 *Hiroshima* (New York: Knopf)

[2] Rhodes R 1988 *The Making of the Atomic Bomb* (New York: Simon & Schuster)

[3] Shizuma K, Iwatani K, Hasai H, Hoshi M, Oka T and Morishima H 1993 Residual $^{152}$Eu and $^{60}$Co induced by neutrons from the Hiroshima atomic bomb *Health Phys.* **65** 272–82

[4] Kato K, Habara M, Aoyama T, Yoshizawa Y, Biebel U, Haberstock G, Heinzl J, Korschinek G, Morinaga H and Nolte E 1988 Measurement of neutron fluence from the Hiroshima atomic bomb *J. Radiat. Res.* **29** 261–6

[5] Straume T, Finkel R C, Eddy D, Kubik P W, Gove H E, Sharma P, Fujita S and Hoshi M 1990 Use of accelerator mass spectrometry in the dosimetry of Hiroshima neutrons *Nucl. Instrum. Methods Phys. Res.* B **52** 552–6

[6] Straume T, Egbert S D, Woolson W A, Finkel R C, Kubik P W, Gove H E, Sharma P and Hoshi M 1992 Discrepancies in the new neutron dosimetry for Hiroshima confirmed using AMS *American Chemical Society, Division of Nuclear Chemistry and Technology* paper 98

[7] Straume T, Egbert S D, Woolson W A, Finkel R C, Kubik P W, Gove H E, Sharma P and Hoshi M 1992 Neutron discrepancies in the DS86 Hiroshima dosimetry system *Health Phys.* **63** 421–6

[8] Straume T 1993 Neutron discrepancies in the DS86 dosimetry system have implications for risk estimates *RERF Update* **4** 3–4 (Hiroshima: Radiation Effects Research Foundation)

## References to Chapter 7

[1] See Peltier R W 1994 Ice Age paleotopography *Science* **265** 195 201 and references therein

[2] Josenhans H, Fedje D, Pienitz R and Southon J 1997 Early humans and rapidly changing Holocene sea levels in the Queen Charlotte Islands–Hecate Strait, British Columbia, Canada *Science* **277** 71–4

[3] Haynes C V Jr 1992 Contributions of radiocarbon dating to the geochronology of the peopling of the New World *Radiocarbon After Four Decades, An Interdisciplinary Perspective* ed R E Taylor, A Long and R S Kra (New York: Springer) pp 355–74

[4]   Kunz M L and Reanier R E 1994 Paleoindians in Beringia: evidence from arctic Alaska *Science* **263** 660–2

[5]   Morell V 1995 Siberia: surprising home for early modern humans *Science* **268** 1279

[6]   King M L and Slobodin S B 1996 A fluted point from the Uptar site, northeastern Siberia *Science* **273** 634–6

[7]   *Rochester Democrat and Chronicle* 1996 Associated Press report 19 August 1996

[8]   Gibbons A 1996 First Americans: not mammoth hunters, but forest dwellers? *Science* **272** 346–7

[9]   Roosevelt A C, da Costa M L, Machado C L, Michab M, Mercier N, Valladas H, Feathers J, Barnett W, da Silveira M I, Henderson A, Sliva J, Chernoff B, Reese D S, Holman J A, Toth N and Schick K 1996 Paleoindian cave dwellers in the Amazon: the peopling of the Americas *Science* **272** 373–84

[10]  Dillehay T D (ed) 1989 *Monte Verde, A Late Pleistocene Settlement in Chile, vol I, Paleoenvironmental and Site Context* (Washington, DC: Smithsonian Institution Press)

[11]  Dillehay T D (ed) 1997 *Monte Verde, A Late Pleistocene Settlement in Chile, vol II, The Archaeological Context* (Washington, DC: Smithsonian Institution Press)

[12]  Gibbons A 1997 Monte Verde: blessed but not confirmed *Science* **275** 1256–7

[13]  Meltzer D J 1997 Monte Verde and the Pleistocene peopling of the Americas *Science* **276** 754–5

[14]  Taylor R E 1992 Radiocarbon dating of bone: to collagen and beyond *Radiocarbon After four Decades, An Interdisciplinary Perspective* ed R E Taylor, A Long and R S Kra (New York: Springer) pp 375–402

[15]  Taylor R E 1983 Non-concordance of radiocarbon and amino acid racemization deduced age estimates on human bone *Radiocarbon* **25** 647–54

[16]  Bada J L, Schroeder R A and Carter G F 1974 New evidence for the antiquity of man in North America deduced from aspartic acid racemization *Science* **184** 791–3

[17]  Bada J L and Helfman P M 1975 Amino acid racemization dating of fossil bones *World Archaeol.* **7** 160–83

[18]  Bada J L, Gillespie R, Gowlett J A J and Hedges R E M 1984 Accelerator mass spectrometry radiocarbon ages of amino acid extracts from California paleoindian skeletons *Nature* **312** 442–4

[19]  Taylor R E, Payen L A, Gerow B, Donahue D J, Zabel T H, Jull A J T and Damon P E 1983 Middle Holocene age of the Sunnyvale human skeleton *Science* **220** 1271–3

[20]  Taylor R E, Payen L A, Prior C A, Slota Jr P J, Gillespie R, Gowlett J A J, Hedges R E M, Jull A J T, Zabel T H, Donahue D J and Berger R 1985 Major revisions in the Pleistocene age assignments for North American human skeletons by [14]C accelerator mass spectrometry: none older than 11,000 [14]C years BP *Am. Antiquity* **50** 136–40

[21]  Egan T 1998 An Indian without reservations *New York Times Magazine* 18 January 1998, pp 16–9

[22]  Gibbons A 1996 The peopling of the Americas *Science* **274** 31–3

[23]  Gibbons A 1996 DNA enters dust-up over old bones *Science* **274** 172

[24]  Jones G D and Harris R J 1977 Contending for the dead *Nature* **386** 15–6

[25]  Gibbons A 1997 Anthropologists 1, Army Corps 0 *Science* **277** 173

[26]  Morell V 1998 Kennewick Man: more bones to pick *Science* **279** 25–6

[27]  Taylor R E private communication

## References to Chapter 8

[1]   Peltier W R 1994 Ice Age paleotopography *Science* **265** 195–201
[2]   Haynes C V 1997 private communication, 4 December 1997
[3]   Conard N, Asch D L, Asch N B, Elmore D, Gove H E, Rubin M, Brown J A, Wiant M D, Farnsworth K B and Cook T G 1984 Accelerator radiocarbon dating of evidence for prehistoric horticulture in Illinois *Nature* **308** 443–6
[4]   Roush W 1997 Squash seeds yield new view of early American farming *Science* **276** 894–5
[5]   Smith B D 1997 The initial domestication of cucurbita pepo in the Americas 10,000 years ago *Science* **276** 932–4
[6]   Saunders J W, Mandel R D, Saucier R T, Allen E T, Hallmark C T, Johnson J K, Jackson E H, Allen C M, Stringer G L, Frink D S, Feathers J K, Williams S, Gremillion K J, Vidrine M F and Jones R 1997 A mound complex in Louisiana at 5400–5000 years before the present *Science* **277** 1796–9
[7]   Pringle H 1997 Oldest mound complex found at Louisiana site *Science* **277** 1761–2
[8]   Mowat F 1990 *West Viking* (Toronto: McClelland & Stewart)
[9]   Ingstad H 1964 Vinland ruins prove Vikings found the New World *National Geographic* **126** 708–34
[10]  Nydal R 1989 A critical review of radiocarbon dating of a Norse settlement at L'Anse aux Meadows, Newfoundland, Canada *Radiocarbon* **31** 976–85
[11]  Litherland A E private communication
[12]  Gove H E 1989 Progress in radiocarbon dating the Shroud of Turin *Radiocarbon* **31** 965–9
[13]  Kaylin J 1996 Tales of the 'Un-Fake' *Yale* May 30–5
[14]  Cahill T private communication
[15]  Buchanan R 1985 SIDELINE, when it came to duck decoys the Paiute Indians made them to last *Sports Illustrated* 25 February 1985


## References to Chapter 9

[1]   1991 (June) *Nuclear Power Plants in the World* 9th edn as of 31 December 1990 (Tokyo: Japan Atomic Industrial Forum)
[2]   1985 (August) *Radioactive Waste, Issues and Answers* (Arvada, CO: American Institute of Professional Geologists)
[3]   1995 (January) *Closing the Circle on the Splitting of the Atom* (US Department of Energy, Office of Environmental Management)
[4]   Beasley T M, Elmore D, Kubik P W and Sharma P 1992 Chlorine-36 releases from the Savannah River site nuclear fuel reprocessing facilities *Ground Water* **30** 539–48
[5]   Elmore D, Tubbs L E, Newman D, Ma X Z, Finkel R, Nishiizumi K, Beer J, Oeschger H and Andree M 1982 $^{36}$Cl bomb pulse measured in a shallow ice core from dye 3, Greenland *Nature* **300** 735–7
[6]   Bentley H W, Phillips F M, Davis S N, Gifford S, Elmore D, Tubbs L E and Gove H E 1982 Thermonuclear $^{36}$Cl pulse in natural water *Nature* **300** 737–40
[7]   Synal H-A, Beer J, Bonani G, Suter M and Woelfli W 1990 Atmospheric transport of bomb-produced $^{36}$Cl *Nucl. Instrum. Methods Phys. Res.* B **52** 483–8

[8] Scanlon B R, Kubik B W, Sharma P, Richter B C and Gove H E 1990 Bomb chlorine-36 analysis in the characterization of unsaturated flow at a proposed radioactive waste disposal facility, Chihuahuan Desert, Texas *Nucl. Instrum. Methods Phys. Res.* B **52** 489–92

[9] Beasley T M, Cecil L D, Sharma P, Kubik P W, Fehn U, Mann L J and Gove H E 1993 Chlorine-36 in the Snake River Plain aquifer at the Idaho National Engineering Laboratory: origins and implications *Ground Water* **31** 302–10

A much more extensive survey of $^{36}$Cl at the Idaho Falls site was carried out later, see Beasley T M 1995 Inventory of site-derived $^{36}$Cl in the Snake River Plain aquifer, Idaho National Engineering Laboratory, Idaho, *Report No EML-567* (New York: Environmental Measurements Laboratory, US Department of Energy)

[10] Kilius L R, Rucklidge J C and Soto C 1994 The dispersal of $^{129}$I from the Columbia River Estuary *Nucl. Instrum. Methods Phys. Res.* B **92** 393–7

[11] Rao U and Fehn U 1996 Application of anthropogenic $^{129}$I as a tracer of nuclear emissions: a study around potential point sources at a nuclear fuel reprocessing plant and two nuclear plants in upstate New York *Nucl. Instrum. Methods Phys. Res.* B **123** 361–6

[12] Norris A E, Wolfsberg K, Gifford S K, Bentley H W and Elmore D 1987 Infiltration at Yucca Mountain, Nevada, traced by $^{36}$Cl *Nucl. Instrum. Methods Phys. Res.* B **29** 376–9

[13] Norris A E, Bentley H W, Cheng S, Kubik P W, Sharma P and Gove H E 1990 $^{36}$Cl studies of water movements deep within unsaturated tuffs *Nucl. Instrum. Methods Phys. Res.* B **52** 455–60

## References to Chapter 10

[1] Elmore D, Gove H E, Beukens R P, Litherland A E, Purser K H and Rubin M 1978 A method for dating the Shroud of Turin *La Sindone e la Scienza, Bilanci e Programmi, Atti del II Congresso Internazionale di Sindonologia* 2nd edn, ed P Coero-Borga (Turin: Edizioni Paoline) pp 428–36

[2] Burleigh R, Leese M and Tite M 1986 An intercomparison of some AMS and small gas counter laboratories *Radiocarbon* **28** 571–7

[3] Gove H E 1987 Turin workshop on radiocarbon dating the Turin Shroud *Nucl. Instrum. Methods Phys. Res.* B **29** 193–5

[4] Gove H E 1989 Progress in radiocarbon dating the Shroud of Turin *Radiocarbon* **31** 965–9

[5] Damon P E, Donahue D J, Gore B H, Hatheway A L, Jull A J T, Linick T W, Sercel P J, Toolin L J, Bronk C R, Hall E T, Hedges R E M, Housley R, Law I A, Perry C, Bonani G, Trumbore S, Woelfli W, Ambers J C, Bowman S G E, Leese M N and Tite M S 1989 Radiocarbon dating of the Shroud of Turin *Nature* **337** 611–5

[6] Gove H E 1990 Dating the Turin Shroud—an assessment *Radiocarbon* **32** 87–92

[7] Phillips T J 1989 Shroud irradiated with neutrons? *Nature* **337** 594

[8] Hedges R E M 1989 Shroud irradiated with neutrons? Hedges replies *Nature* **337** 594

[9] Kouznetsov D A, Ivanov A A and Veletksy P R 1996 Effects of fires and biofractionation of carbon isotopes on results of radiocarbon dating of old textiles: the Shroud of Turin *J. Archaeol. Sci.* **23** 109–21

[10] Jull A J T, Donahue D J and Damon P E 1996 Factors affecting the apparent radiocarbon age of textiles: a comment on 'effects of fires and

biofractionation of carbon isotopes on results of radiocarbon dating of old textiles: the Shroud of Turin', by D A Kouznetsov *et al J. Archaeol. Sci.* **23** 157–60

[11] Burleigh R, Ambers J and Matthews K 1982 *British Museum Natural Radiocarbon Measurements XV; Radiocarbon* **24** 262–90

[12] Gove H E, Mattingly S J, David A R and Garza-Valdes L A 1997 A problematic source of organic contamination of linen *Nucl. Instrum. Methods Phys. Res.* B **123** 504–7

## References to Chapter 11

[1] Roberts D, Garrett K and Harlin G 1993 The Iceman: lone voyager from the copper age *National Geographic* **183** 36–67

[2] Prinoth-Fornwagner R and Nicklaus Th R 1994 The man in the ice: results from radiocarbon dating *Nucl. Instrum. Methods Phys Res.* B **92** 282–90

[3] Spindler K 1994 The Iceman's last weeks *Nucl. Instrum. Methods Phys. Res.* B **92** 274–81

[4] LaMarche V C Jr 1978 Application of the new radiocarbon technique in terrestrial paleoclimatology and paleobiology *Proc. First Conf. on Radiocarbon Dating with Accelerators (University of Rochester)* ed H E Gove pp 314–9

[5] Bard E, Hamelin B, Fairbanks R G and Zindler A 1990 Calibration of the $^{14}C$ timescale over the past 30,000 years using mass spectrometric U–Th ages from Barbados coral *Nature* **345** 405–10

[6] Keilson J and Waterhouse C 1978 Possible impact of the new spectrometric techniques on $^{14}C$ tracer kinetic studies in medicine *Proc. First Conf. on Radiocarbon Dating with Accelerators (University of Rochester)* ed H E Gove pp 391–7

[7] Elmore D 1987 Ultrasensitive radioisotope, stable isotope and trace-element analysis in the biological sciences using tandem accelerator mass spectrometry *Biol. Trace Elem. Res.* **12** 231–45

[8] Vogel J S, Turteltaub K W, Felton J S, Gledhill B L, Nelson D E, Southon J R, Proctor I D and Davis J C 1990 Application of AMS to the biomedical sciences *Nucl. Instrum. Methods Phys. Res.* B **52** 524 30

[9] Felton J S, Turteltaub K W, Vogel J S, Balhorn R, Gledhill B L, Southon J R, Caffee M W, Finkel R C, Nelson D E, Proctor I D and Davis J C 1990 Accelerator mass spectrometry in the biomedical sciences: applications in low-exposure biomedical and environmental dosimetry *Nucl. Instrum. Methods Phys. Res.* B **52** 517–23

[10] Turteltaub K W, Felton J S, Gledhill B L, Vogel J S, Southon J R, Caffee M W, Finkel R C, Nelson D E, Proctor I D and Davis J C 1990 Accelerator mass spectrometry in biomedical dosimetry: relationship between low-level exposure and covalent binding of heterocyclic amine carcinogens to DNA *Proc. Natl. Acad. Sci. USA* **87** 5288–92

[11] Elmore D, Bhattacharyya M H, Sacco-Gibson N and Peterson D P 1990 Calcium-41 as a long-term biological tracer for bone resorbtion *Nucl. Instrum. Methods Phys. Res.* B **52** 531–5

[12] Johnson R R, Berkovits D, Boaretto E, Gelbart Z, Ghelberg S, Meirav O, Paul M, Prior J, Sossi V and Venczel E 1994 Calcium resorbtion from bone in a human studied by $^{41}Ca$ tracing *Nucl. Instrum. Methods Phys. Res.* B **92** 483–8

[13] Meirav O, Sutton R A L, Fink D, Middleton R, Klein J, Walker V R, Halabe A, Vetterli D and Johnson R R 1990 Application of accelerator mass

spectrometry in aluminum metabolism studies *Nucl. Instrum. Methods Phys. Res.* B **52** 536–9

[14]   Barker J, Day J P, Aitken T W, Charlesworth T R, Cunningham R C, Drumm P V, Lilley J S, Newton G W A and Smithson M J 1990 Development of [26]Al accelerator mass spectrometry for biological and toxicological applications *Nucl. Instrum. Methods Phys. Res.* B **52** 540–3

[15]   Johnson R R *et al* The study of aluminum kinetics in healthy and uremic human patients (private communication)

[16]   Edwardson J A, Moore P B, Ferrier I N, Lilley J S, Newton G W, Barker J, Templar J and Day J P 1993 Effect of silicon on gastrointestinal absorption of aluminum *The Lancet* **342** 211–2

[17]   Flarend R and Elmore D 1997 Aluminum-26 as a biological tracer using accelerator mass spectrometry *Aluminum toxicity in Infant's Health and Disease* ed P Zatta and A C Alfrey (Singapore: World Scientific) Chapter 2

[18]   Freeman S P H T, Bronk C R and Hedges R E M 1990 The design of a radiocarbon microprobe for tracer mapping in biological specimens *Nucl. Instrum. Methods Phys. Res.* B **52** 405–9

[19]   Roberts M L, Velsko C and Turteltaub K W 1994 Tritium AMS for biomedical applications *Nucl. Instrum. Methods Phys. Res.* B **92** 459–62

[20]   Vogel J S and Turteltaub K W 1994 Accelerator mass spectrometry in biomedical research *Nucl. Instrum. Methods Phys. Res.* B **92** 445–53

[21]   Purser K H and Gove H E 1990 A new instrument for ultra-sensitive [14]C tracers *Trans. Am. Nucl. Soc.* **62** 10–1

[22]   Purser K H 1994 A future AMS/chromatography instrument for biochemical and environmental measurements *Nucl. Instrum. Methods Phys. Res.* B **92** 201–6

[23]   Bentley H W, Phillips F M, Davis S N, Habermehl M A, Airey P L, Calf G E, Elmore D, Gove H E and Torgersen T 1986 Chlorine-36 dating of very old groundwater 1. The Great Artesian Basin, Australia *Water Resources Res.* **22** 1991–2001

[24]   Torgeson T, Habermehl M A, Phillips F M, Elmore D, Kubik P, Jones B G, Hemmick T and Gove H E 1991 Chlorine-36 dating of very old groundwater 3. Further studies in the Great Artesian Basin, Australia *Waters Resources Res.* **27** 3201–13

[25]   Gove H E 1987 Tandem-accelerator mass-spectrometry measurements of [36]Cl, [129]I and osmium isotopes in diverse natural samples *Phil. Trans. R. Soc.* A **323** 103–19

[26]   Phillips F M, Zreda M G, Smith S S, Elmore D, Kubik P W, Dorn I and Roddy D J 1991 Age and geomorphic history of Meteor Crater, Arizona, from cosmogenic [36]Cl and [14]C in rock varnish *Geochim. Cosmochim. Acta* **55** 2695

[27]   There are many books on the Dead Sea scrolls. One of the early ones is Burrows M 1956 *The Dead Sea Scrolls* (New York: Viking)

[28]   Bonani G, Broshi M, Carmi I, Ivy S, Strugnell J and Woefli W 1991 Radiocarbon dating of the Dead Sea scrolls *Atiqot* **20** 27–32

[29]   Bonani G, Ivy S, Woelfli W, Broshi M, Carmi I and Strugnell J 1992 Radiocarbon dating of fourteen Dead Sea scrolls *Radiocarbon* **34** 843–9

[30]   Jull A J T, Donahue D J, Broshi M and Tov E 1995 Radiocarbon dating of scrolls and linen fragments from the Judean Desert *Radiocarbon* **37** 11–9

# References to Chapter 12

[1] See [6–20] in Chapter 11
[2] See [21] and [22] in Chapter 11
[3] Anthony J M and Donahue D J 1987 Accelerator mass spectrometry solutions to semiconductor problems *Nucl. Instrum. Methods Phys. Res.* B **29** 77–82
[4] Hossain T Z, Elmore D, Gove H E, Hemmick T K, Kubik P W and Jiang S 1987 Neutron activation analysis/accelerator mass spectrometry measurements of nitrogen and chlorine in silicon *Workshop on Applications of Nuclear Physics Techniques to Condensed Matter Physics (American Physical Society, Rutgers, NJ, 14 October)* (unpublished)
[5] Datar S A, Gove H E, Teng R T D and Lavine J P 1995 AMS studies of the diffusion of chlorine in silicon wafers *Nucl. Instrum. Methods Phys. Res.* B **99** 549–52
[6] Gove H E, Kubik P W, Sharma P, Datar S, Fehn U, Hossain T Z, Koffer J, Lavine J P, Lee S-T and Elmore D 1990 Applications of AMS to electronic and silver halide imaging research *Nucl. Instrum. Methods Phys. Res.* B **52** 502–6
[7] Bronk C R and Hedges R E M 1987 A gas ion source for radiocarbon dating *Nucl. Instrum. Methods Phys. Res.* B **29** 45–9
[8] Ramsey C B and Hedges R E M 1997 Hybrid ion sources: radiocarbon measurements from microgram to milligram *Nucl. Instrum. Methods Phys. Res.* B **123** 539–45
[9] Wills J S C, Schmeing H, Diamond W T, Diserens J, Imahari Y and Taylor T 1996 A microwave-driven negative ion source *Rev. Sci. Instrum.* **67** 1227–9
[10] Schneider R J, von Reden K F, Wills J S C, Diamond W T, Lewis R, Savard G and Schmeing H 1997 Hold-up and memory effect for carbon in a compact microwave ion source *Nucl. Instrum. Methods Phys. Res.* B **123** 546–9
[11] Rucklidge J C, Gorton M P, Wilson G C, Kilius L R, Litherland A E, Elmore D and Gove H E 1982 Measurement of Pt and Ir at sub-ppb levels using tandem accelerator mass spectrometry *Can. Mineral.* **20** 111–9
Rucklidge J C, Gorton M P, Wilson G C, de Gasparis S, Kilius L R, Litherland A E, Elmore D and Gove H E 1982 Accelerator mass spectrometric (AMS) measurements of Pt and Ir at the Cretaceous–Tertiary boundary *Syllogeus (National Museums of Canada)* **39** 147–9
See also [5] in Chapter 5
[12] Kilius L R, Baba N, Garwan M A, Litherland A E, Nadeau M-J, Rucklidge J C, Wilson G C and Zhao X-L 1990 AMS of heavy ions with small accelerators *Nucl. Instrum. Methods Phys. Res.* B **52** 357–65
[13] Zhao X-L, Nadeau M-J, Kilius L R and Litherland A E 1994 The first detection of naturally-occurring $^{236}$U with accelerator mass spectrometry *Nucl. Instrum. Methods Phys. Res.* B **92** 249–53
[14] Fifield L K, Cresswell R G, di Tada M L, Ophel T R, Day J P, Clacher A P, King S J and Priest N D 1996 Accelerator mass spectrometry of plutonium isotopes *Nucl. Instrum. Methods Phys. Res.* B **117** 295–303
[15] Fifield L K, Clacher A P, Morris K, King S J, Cresswell R G, Day J P and Livens F R 1997 Accelerator mass spectrometry of the planetary elements *Nucl. Instrum. Methods Phys. Res.* B **123** 400–4
[16] Kutschera W 1988 Present and future prospects of accelerator mass spectrometry *Nucl. Instrum. Methods* A **268** 552–60
[17] See [8] in Chapter 4
[18] Purser K H, Litherland A E and Gove H E 1979 Ultra-sensitive particle

identification systems based upon electrostatic accelerators *Nucl. Instrum. Methods* **162** 637–56

[19] Litherland A E, and Kilius L R 1997 Neutral injection for AMS *Nucl. Instrum. Methods Phys. Res.* B **123** 18–21

[20] Betz H D 1972 Charge states and charge-changing cross sections of fast heavy ions *Rev. Mod. Phys.* **44** 465–539

[21] Wittkower A B and Betz H D 1973 Equilibrium charge state distributions of energetic ions ($Z > 2$) in gaseous and solid media *Atomic Data* **5** 113–66

[22] Wittkower A B and Betz H D 1973 Equilibrium charge state distributions of 2–15 MeV tantalum and uranium ions stripped in gases and solids *Phys. Rev.* A **7** 159–67

[23] Gove H E 1987 Tandem-accelerator mass-spectrometry measurements of $^{36}$Cl, $^{129}$I and osmium isotopes in diverse natural samples *Phil. Trans. R. Soc.* A **323** 103–19

[24] Rucklidge J C, Evensen N M, Gorton M P, Beukens R P, Kilius L R, Lee H W, Litherland A E, Elmore D, Gove H E and Purser K H 1981 Rare isotope detection with tandem accelerators *Nucl. Instrum. Methods* **191** 1–9

[25] Wilson G C and Rucklidge J R 1990 Geochemistry and mineralogy of the Late Archean Owl Creek shear-hosted gold deposit Timmins, Ontario, Canada, *Third Int. Archaean Symp. (Perth, Western Australia, September 1990)* extended abstracts volume, pp 391–3 (unpublished)

[26] Wilson G C, Rucklidge J R and Kilius L R 1990 Sulfide gold content of Skarn mineralization at Rossland, British Columbia *Econ. Geol.* **85** 1252–9

[27] Wilson G C, Kilius L R and Rucklidge J C 1991 *In situ* analysis of precious metals in polished mineral samples and sulfide 'standards' by accelerator mass spectrometry at concentrations of parts-per-billion *Geochim. Cosmochim. Acta* **55** 2241–51

[28] Rucklidge J C, Wilson G C, Kilius L R and Cabri L J 1992 Trace element analysis of sulfide concentrates from Sudbury by accelerator mass spectrometry *Can. Mineral.* **30** 1023–32

[29] Li C, Naldrett A J, Rucklidge J C and Kilius L R 1993 Concentrations of platinum group elements and gold in sulfides from the Strathcona deposit, Sudbury, Ontario *Can. Mineral.* **30** 523–31

[30] Chai G, Naldrett A J, Rucklidge J C and Kilius L R 1993 *In situ* qualitative analysis for PGE and Au in sulfide minerals of Jinchuan Ni–Cu deposit by accelerator mass spectrometry *Can. Mineral.* **31** 19–30

[31] Rucklidge J C 1995 Accelerator mass spectrometry in environmental geoscience *Analyst* **120** 1283–90

[32] Wilson G C, Rucklidge J C and Kilius L R 1995 Ultrasensitive trace element analysis with accelerator mass spectrometry: the current state of the art *Can. Mineral.* **33** 237–42

[33] Wilson G C, Kilius L R and Rucklidge J C 1995 Precious metal contents of sulfide, oxide and graphite crystals: determinations by accelerator mass spectrometry *Econ. Geol.* **90** 255–70

[34] Wilson G C, Kilius L R, Rucklidge J C, Ding G-J and Zhao X-L 1997 Trace-element analysis of mineral grains using accelerator mass spectrometry— from sampling to interpretation *Nucl. Instrum. Methods Phys. Res.* B **123** 579–82

[35] Wilson G C, Pavlish L A, Ding G-J and Farquhar R M 1997 Textural and *in-situ* analytical constraints on the provenance of smelted and native archaeological copper in the Great Lakes Region of eastern North America *Nucl. Instrum. Methods Phys. Res.* B **123** 498–503

[36] Ding G-J, Kilius L R, Wilson G C, Zhao X-L and Rucklidge J C 1997 *In-situ* AMS determinations of Re–Os isochron in IIA iron meteorites *Nucl.*

*Instrum. Methods Phys. Res.* B **123** 424–30

[37] Ding G-J, Kilius L R, Wilson G C, Zhao X-L and Rucklidge J C 1997 Evidence for anomalous [107]Ag and [109]Ag compositions in iron meteorites *Nucl. Instrum. Methods Phys. Res.* B **123** 414–23

[38] Wilson G C, Rucklidge J C, Kilius L R, Ding G-J and Cresswell R G 1997 Precious metal abundances in selected iron meteorites: *in-situ* AMS measurements of the six platinum group elements plus gold *Nucl. Instrum. Methods Phys. Res.* B **123** 583–8

[39] Krestow J S A 1997 A new analytical technique for measuring platinum group element concentrations in insulating samples *PhD Thesis Proposal* Department of Geology, University of Toronto

[40] Naldrett A J and Barnes S-J 1986 The behaviour of platinum group elements during fractional crystallization and partial melting with special reference to the composition of magmatic sulfide ores *Forsch. Mineral.* **64** 113–33

[41] A J Naldrett private communication

[42] Carrigan C R, Heinle R A, Hudson G B, Nitao J J and Zucca J J 1996 Trace gas emissions on geological faults as indicators of underground nuclear testing *Nature* **382** 528–31

[43] Litherland A E and Purser K H private communication

# Index